ANDREY BOGINSKY

METHODOLOGICAL BASIS
FOR CREATING PROMISING
HIGH-TECH PRODUCTS
IN THE DIGITAL TRANSFORMATION

Monograph

EurAsian
Scientific Editions Ltd

Tallinn, 2021

UDC 330.101.542:004
LBC 65.012.1я73
B 71

ISBN 978-9949-7485-4-9

Andrey Boginsky. Methodological basis for creating promising high-tech products in the digital transformation. – Talinn: EurAsian Scientific Editions Ltd, 2021 – 252 p.

ISBN 978-9949-7485-4-9

www.eurasian-scientific-editions.org

Cover photo credit: Fabio on Unsplash

INTRODUCTION

In the current conditions of economic globalization and the redistribution of spheres of economic influence in the world, the development of industries is closely related to ensuring the competitiveness of their products in the domestic and global markets, based on a permanent innovation process, the development and implementation of new modern technologies and equipment, as well as improving management methods, due to the need to meet the requirements of digital transformation of all spheres of society.

To solve the urgent task of ensuring sustainable economic growth in the context of digital transformation and the rapid development of science and technology, organizations need to adapt the principles of conducting their economic activities to the new realities, as well as radically restructure the design and production processes focused on creating products with new consumer properties with exceptional value to the consumer.

The development of innovations and the emergence of new technologies in the modern world are becoming the main factor in increasing competitiveness in the international arena, including in overcoming crisis situations and creating conditions for the formation of new consumer markets. In this context, a special role is played by breakthrough innovations, the pace of development of which corresponds to the pace of development of technological structures, within which there is an accumulation of unique technological and organizational competencies.

The development of new technological and organizational competencies leads to an increase in the intellectual potential of organizations and society as a whole, which affects the reduction of the time for changing technological patterns, leading to an intellectual explosion and the emergence of new waves of innovation.

The transformational shifts of the economy are greatly influenced by the fourth industrial revolution or Industry 4.0, as a result of which digital manufacturing and smart enterprises are created and actively developed, characterized by a high ability to transform and adapt to changing consumer expectations, resource efficiency and ergonomics, as well as the integration of all subjects of economic relations (consumers, investors, partners, etc.) into business and value creation processes. This, in turn, stimulates the creation of unique highly competitive products that can meet the new needs of society or form new consumer markets, ensuring the advanced development of industrial enterprises. This process should be based on the implementation of breakthrough innovative ideas and technologies that surpass the solutions available on the market in terms of their technological level. The implementation of such innovative ideas in practice creates the foundation for sustainable economic development.

Considering the issues of sustainable economic development of an organization and management of the processes of creating promising high-tech products, basic principles of managing the development of new products are being developed, key technological competencies are considered to ensure the creation of promising competitive products, and methods are proposed for assessing the attainability of techno-economic indicators of products through the use of innovative technologies and competencies in the implementation of innovative projects.

Since in the course of the implementation of innovative projects of high-tech production, the economic and technological effect is achieved as a result of the implementation of various organizational and technical measures with varying degrees of efficiency, special attention should be paid to the study of economic tools for planning resource support for the life cycle of promising high-tech products, methods for planning the cost of R&d and an economic and mathematical model for the optimal planning of organizational and technical measures for the implementation of innovative projects in conditions of resource constraints, as well as cost optimization at various stages of the life cycle of innovative projects to create high-tech products.

In this regard, the impact of modern digital technologies on the product value chain is assessed, the role of digital models and digital modeling in cost optimization at the stages of the product life cycle is studied.

All the developed economic tools for managing the process of creating unique highly competitive products through their integrated use are reduced to the developed economic mechanism for managing the competitiveness of promising high-tech products, taking into account the use of digital technologies of development and production in order to optimize production and technological processes in order to ensure the competitiveness of high-tech organizations and their economic stability, including by updating products.

As part of the study of product renewal as a mechanism for maintaining global competitive leadership, the growth of the volume of own processing during product renewal is investigated as a source of creating competitive advantages of the organization, and a method is proposed for a comprehensive assessment of the effectiveness of the product renewal process when it is created for the specified technical and economic parameters.

The above-described methods, economic tools and mechanism allowed us to form a system for managing the manufacture of new products in the context of digital transformation with an assessment of its competitiveness, as well as to offer the basis for creating global competitive advantage and an algorithm for forming business strategies that allow us to find a way to achieve competitive advantages over the products and services of competitors in the market, which ensures stable economic development of the company.

The monograph will be of interest to scientists, theoretical economists dealing with the problems of microeconomics, and industrial practitioners, heads of industrial organizations, as well as researchers interested in the described issues.

CHAPTER 1. BASIS FOR ENSURING THE COMPETITIVENESS OF HIGH-TECH PRODUCTS IN THE DIGITAL AGE

1.1. BASIS FOR SUSTAINABLE ECONOMIC DEVELOPMENT OF ORGANIZATIONS IN THE CONTEXT OF DIGITAL TRANSFORMATION

In the context of rapid scientific, technical and technological progress that lead to the transformation of the principles of economic activity, as well as the design and manufacture of products with new consumer properties that are of exceptional value to the consumer, organizations face an urgent task to ensure their sustainable economic growth, especially in the context of digital transformation. The global market as a whole, as well as developed and developing countries, is characterized by a change in the structure of the economy under the influence of cyber-physical and cyber-economic systems and technologies in the context of a change in waves of innovations, each with its unique breakthrough innovations. At the same time, the periods of waves of innovations are gradually decreasing as a result of the growth of the intellectual potential of companies and society as a whole on the basis of the development of new technological and organizational competencies, which eventually leads to an intelligence explosion and the emergence of a new wave of innovation.

Currently, in the industry of different countries, the technologies of the fourth, fifth and sixth waves of innovation are mainly present in various ratios, characterized by the active development of space technologies, nanoelectronics and robotics. Experts expect the seventh wave of innovation. B. Zebuhr in his article "The seventh wave and beyond: a world revolution driven by knowledge"[1]

---

[1] Zebuhr B. The seventh wave and beyond: a world revolution driven by knowledge // Infinite Energy. 2008. Issue 78. P. 9. The text is found in the Internet: https://www.infinite-energy.com/iemagazine/issue78/theseventhwaveeditorial.html.

noted the following: "I believe that the seventh wave will become a super wave of innovation and active development of new technologies based on the six preceding waves." He emphasized that super wave means a wave of innovations based on a fundamentally new science, characterized by the dominance of a range of different technologies. Thus, according to experts, the technologies of the seventh wave will lead to the active development of knowledge-based industries and artificial intelligence.

At the same time, the ongoing fourth industrial revolution, otherwise called Industry 4.0, is causing such transformational shifts as the formation of digital manufacturing, an intelligent enterprise characterized by high capacity for change and adaptation to changing consumer expectations, resource efficiency and ergonomics as well as integration of all economic actors (consumers, investors, partners, etc.) to business processes and value creation. The use of Industry 4.0 technologies determines the transition of organizations to the digital enterprise level. However, an important task arises to ensure the advanced development of industrial enterprises through creating unique highly competitive products that can meet the new needs of society or form new consumer markets. As a rule, the advanced development of organizations is based on the implementation of breakthrough innovative ideas and technologies that exceed the technological level of the solutions available on the market.

Some of the basic technologies of Industry 4.0 are big data processing and analysis technology, the Internet of things (IoT) (5G), virtual and augmented reality, etc. The essence of big data technology is to efficiently process and extract knowledge from a huge array of structured and unstructured data using scalable software developed in the early 2000s as an alternative to traditional database management systems.

In addition to the listed basic technologies, digital enterprises are also characterized by the presence of:

- supercomputer technologies that include a set of tools that surpass most existing computers in the world in terms of their technical parameters and computing speed. These technologies can be used to solve specialized problems of digital enterprises;

- additive technologies (3D printing), implying a universal method of creating physical objects and their parts on the basis of a single platform, which allows a single production approach implemented in digital form;
- technology of the Internet of things, which allows to ensure the flexibility of production, avoiding rigid "conveyor" solutions, to increase the level of control over production thanks to a single technological platform, as well as to increase production efficiency by reducing costs associated with the human factor (errors, downtime, high cost of human labor);
- technologies of virtual and augmented reality, which makes it possible to construct virtual objects using specialized software and programs (for example, a digital twin of products);
- quantum computing technologies, including the recording, storage and transmission of digital data based on physical systems;
- autonomous robot technology, which involves the transfer of robots performing complex production functions to autonomous operation thanks to 5G technologies. The use and development of this technology can subsequently lead to joint work of robots with people, as well as to the functioning of drones and machines on autopilot.

In Industry 4.0, the key drivers of economic activity are electronic technologies and services, as well as the digitalization of large amounts of data, the processing and analysis of which can significantly improve efficiency and quality in the production and consumption of goods, works and services, as well as in enterprise management procedures. Figure 1.1 shows a diagram of the organization of the production process in a digital enterprise.

At the stage of digital design and building a digital twin of a product, enterprises seeking to achieve advanced development should use modern digital intellectual methods for designing and modeling products based on technologies for building digital twins

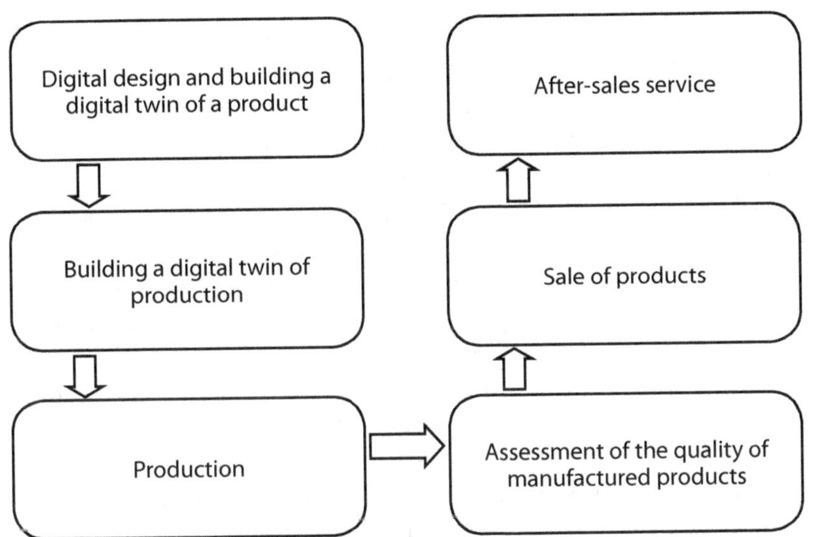

Figure 1.1. Organization of the production process in a digital enterprise

of products and full-fledged mathematical modeling with simulation of various scenarios of product functioning in order to design its image that corresponds fairly well to the specified technical characteristics of products.

In order to successfully design and model a high-tech product, enterprises seeking to achieve advanced development should fully apply machine learning and artificial intelligence technologies, which use big data mining to find the most effective design solutions.

For the pre-production of complex products, a digital twin of the product and a digital production plan are used, which create a single virtual environment for modeling production processes taking into account various factors and scenario variations. In organizations with advanced development and creation of a digital product twin, the pre-production stage also includes its robotization – equipping production with intelligent robotic systems that are able to quickly respond to changes in the work area and adapt their activities to such changes. When organizing production processes, cybernetic systems are used, consisting of natural and artificially created objects, subsystems and controllers, which form a functionally improved automated control system.

At the manufacturing stage, an important role is played by automated and robotic complexes that not only perform mechanical work, but also independently plan and manage other robotic complexes without human intervention. Based on the latest achievements of artificial intelligence, such complexes can be delegated not only production processes, but also "machine management" processes that will be able to predict and solve production issues more accurately and efficiently than humans.

At the same time, a number of critical production sites can use the "Brain-computer interface" technology, which will allow for effective interaction between a person and control complexes to adjust production processes in real time. Currently, there is an example of interaction between a human and a robotic complex at the physical level – an exoskeleton. The main purpose of this suit, invented for workers who perform physical labor – is to increase the strength of human muscles, expand the range of movements, and reduce the maximum physical load on the muscles and bones of the worker. It is assumed that the use of this suit will significantly increase productivity and the level of labor safety, reduce the degree of human fatigue and, accordingly, enhance his motivation and job satisfaction.

In addition, augmented reality technologies are very promising for enterprises of the future, when a highly qualified employee can perceive the production process using digital and virtual technologies.

In terms of manufacturing products and equipment for these purposes, additive technologies must be used, which consist in the creation of products by sequentially applying layers. Models made by the additive method can be used both at the stage of pilot production and mass production. Additive methods (for example, welding) have been used in manufacturing for a long time. However, in the enterprises of the future, they must be based on the most innovative three-dimensional ways of creating a product. For example, 3D printing, in which an object consists of layers that are superimposed on each other using a 3D printer.

To solve the problems of assessing the quality of manufactured products and their after-sales service, it is also proposed to use feedback technologies during operation: the Internet of Things (IoT) and mobile applications and devices (mobile technologies). The Internet of things refers to the ability of a system to automatically make routine decisions due to the "communication" of things in the system, which implies their ability to identify, transmit data to each other and process it. Thus, at the stage of after-sales service of products, a decision can be automatically made about the need to replace parts, repair, etc. Morever, mobile applications can monitor the status of products and inform owners about breakdowns and the need for a scheduled inspection of the product by specialists, offering a list of the nearest organizations where they can receive assistance. It is logical to assume that the widespread use of the described technologies leads to the transformation of the traditional chain "economy – production – economy" as a form of creating competitive products. This transformation consists in the use of Industry 4.0 technologies both in shaping the techno-economic image of promising products (forecasting the future needs of organizations and society as a whole based on the analysis of big data in the global information space), and in identifying ways to achieve price and non-price competitive advantages of products (for example, the use of digital design and modeling methods, design at a given cost and competitiveness using digital twins of products, organizations, competitive market environment). In the conditions of mutual convergence of manufacturers and buyers in the market as a result of digitalization of their interaction, the optimal tools for achieving high competitive advantages are personification and customization of products, which consists in the manufacture of products with techno-economic characteristics that meet the needs of a particular consumer at a high level. It can be achieved, on the one hand, through the use of flexible production systems and digital transformation of the main technological processes, and on the other hand, through closer communication between consumers and manufacturers in the global information space.

The described technologies and the need for their application set the following tasks for a modern organization:

- ability to maintain the competitiveness of products at a high level;
- mass introduction of intelligent (quantum) sensors into equipment and production lines;
- transition to unmanned production and mass introduction of robotic technologies;
- transition to storage of information and computing capacity to distributed resources (a cloud-based technology);
- end-to-end automation and integration of production and management processes into a single information system;
- digitization of all technological documentation, creation of an electronic document management system;
- digital design and modeling of technological processes of an organization and products throughout the entire life cycle;
- use of material augmentation technologies (additive technologies);
- use of mobile technologies for monitoring and controlling processes in organizations and after-sales service of manufactured products;
- forecasting the quality of products.

The ability to maintain the competitiveness of products at a high level, which is closely related to the sustainable economic development of an organization, is a particularly urgent task. At the same time, sustainable economic development is impossible without effective solution of the following set of economic problems (Fig. 1.2).

A sustainably developing organization is able to generate the resources necessary to build new innovative technologies and key competencies, which ultimately determines the strong market position of the organization.

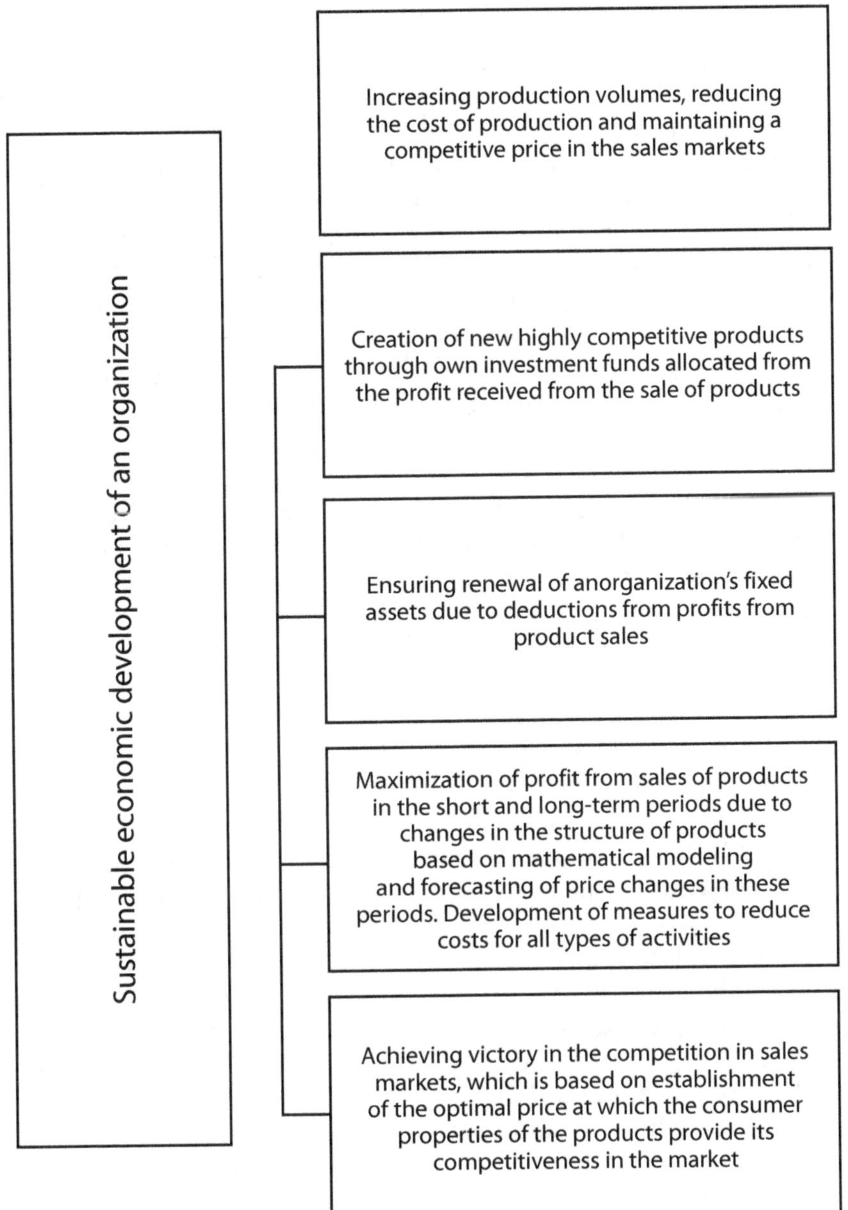

Figure: 1.2. Components of sustainable development of an organization

At the same time, in the conditions of the dominance of the service sector over the traditional manufacturing sector, many manufacturers of high-tech products are forced to look for new business models for their sale. This is due to the fact that such products are not always in high demand among the mass consumer (for example, due to their specificity or high cost). The products of highly specialized knowledge-based organizations cannot be sold on a "free market", since it does not provide many buyers and suppliers, and competition among organizations in the industry is not able to reduce the cost of products towards stable, cost-effective, equilibrium prices[2]. The state or other companies are the traditional consumers of such products. An organization's transition to the mass-market segment can be related to the distribution of product characteristics that have consumer value through service sales models. For example, a helicopter company can sell on the market only a limited number of machines to the state or large companies, but a helicopter manufacturer can enter the mass-market segment through the sale of a flight hour. Moreover, modern digital technologies allow for real-time communication between the manufacturer (owner) of the helicopter and the consumer of the corresponding service for a one-time helicopter flight.

Thus, through the personalization of products and the use of new business models for the distribution of products and services, a company's product range can be expanded, and the products can have new competitive advantages.

In high-tech organizations that manufacture products for both the state customer and the commercial market, the expansion of the product line is usually closely linked to the need to produce competitive products for the competitive market in the face of the reduced volume of state orders. The most important systemic process that ensures the sustainable economic development of an organization is the diversification of the manufactured product range. The purpose of such diversification is to manage the process of achieving a sustainable economic state and its further maintenance by timely

---

2 Yu. V. Vlasov, A. A. Chursin, Management of Diversificataion System in Aerospace Industry // Economy of region. – 2016. – V. 12. – No. 4. – Pp. 1205–1217.

updating products, taking into account consumer demand and creating value for the buyer. Depending on the specifics of organizations, the goals of diversification can be specified.

A wide range of products and services provided by diversified companies creates conditions for maintaining the pace of sustainable economic development, as well as makes it possible to transfer internal capital, technologies and competencies from one production sector to another, ensuring intensive innovative development.

The most important issue in the implementation of measures to expand the product line is their resource provision. The resources required to expand the types of goods produced are shown in Fig. 1.3.

Note that the sustainable development of high-tech organizations can be promoted by an effective industrial policy implemented by the state in the following areas.

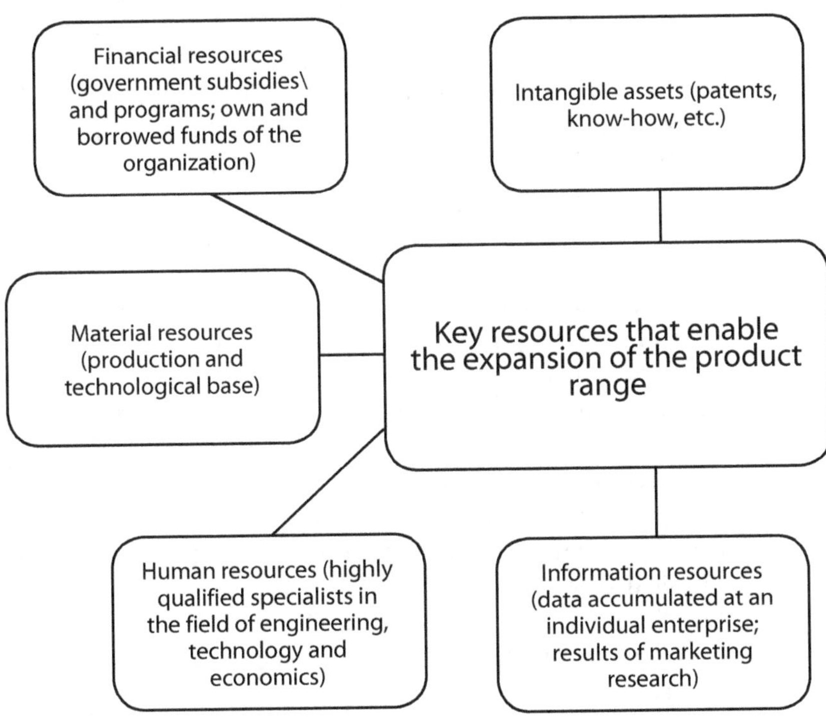

Figure: 1.3. Resource support for expanding the product range

1. Adoption and implementation of a set of measures to improve state regulation mechanisms that contribute to the intensive development and implementation of innovative solutions by organizations in order to ensure the competitiveness of products with high added value for their placing on world markets and implementation of import substitution programs, creation of technological independence, etc. Measures are implied for:
   - development of a system of coordination and management of knowledge-based industries;
   - improving the system of state regulation and support of knowledge-based industries that produce goods that are competitive on world markets;
   - improving the methodology for predicting the level of competitiveness of an organization, its products, and the knowledge-based industry as a whole;
   - development of legislative advantages (tax holidays, financial incentives, etc.) for organizations introducing innovative technologies into production, and other measures of state protectionism for the country's transition to an innovative path of development and increasing competitiveness.

2. Implementation of state (public-private) programs and projects for the development and production of industrial innovative high-tech products that meet the latest achievements in science, technology and technology (taking into account the modernization of production, completely new equipment, support for research and experimental design developments in creating breakthrough technologies, including those based on new physical principles).

3. Development of approaches and measures for state incentives for investment in various sectors of the economy to ensure sustainable innovative development of organizations. Creation of backbone elements of the Russian innovation system and mechanisms for attracting funds from

private sources to the innovation sphere of Russia to activate business through various state tools for regulating and managing the economy[3].

Note that in managing the competitiveness of a product portfolio, a high-tech organization must maintain a balance between expansion and concentration of production, since it has the ability to concentrate all its resources on a specific area and achieve significant competitive advantages in it. In this case, the competitive position of an organization will be protected by high barriers from the entry of competitors' products into the market and the same high requirements for the goods produced[4].

In terms of resource provision, the expansion of the product range of an organization is expressed in the distribution of resources among several types of activities. Since the functioning of an organization is always fraught with the risk of loss, the expansion of the product range has a positive effect related to the reduction of possible losses per type of product.

As a quantitative indicator characterizing the change in the economic stability of an organization due to the expansion of its product line, let us consider the operating profit of NOP. This indicator is calculated as follows:

$$NOP = TR - VC - FC,$$

where TR is total cost revenue from product sales; VC is variable cost; FC – is fixed production cost.

Fixed costs practically do not change with increasing output. Variable costs are not adjusted per unit of output, but are changed by total output, in proportion to its volume[5].

---

3 Chursin A.A., Ivanov A.S. Recommendations for creating a concept for sustainable innovative development of structures with consolidated capital //Defense complex – scientific and technical progress of Russia. – 2010. – No. 3. – Pp. 110-117.

4 Borisoglebskaya L. N., Mukhin I. E., Evtushenko O. N. Methods and mechanisms for ensuring the competitiveness of defense industry products in modern conditions // Bulletin of the Academy of military Sciences. – 2016. – № 1 (54). – Pp. 104-113.

5 Evtushenko O. N. Diversification as a tool for optimizing the activities of integrated structures. Horizons of Economics. – 2017. – № 3 (36). – Pp. 59-63.

As a result of the expansion of the product line, the volume of product sales is adjusted. The operating leverage (*DOL*) allows quantifying the change in profit volumes depending on changes in sales volumes and is calculated as follows, presented in the paper[6]:

$$DOL = \frac{TR - VC}{TR - VC - FC}.$$

The operating leverage *DOL* shows the percentage of an organization's operating profit change when revenue changes by 1%.

Let us consider the application of the described mathematical model on a specific example. We will assume that the expansion of the product line is carried out in six stages, each of which produces new types of products that have competitive advantages (table 1.1).

Table 1.1. Example of evaluating product line expansion (in standard units)

Stage	Sales volume	Revenue TR	VC	FC	NOP, %	DOL
0	70	420	280	100	9,52	3,5
1	80	480	320	100	12,5	2,7
2	90	540	360	100	14,81	2,25
3	100	600	400	100	16,67	2
4	110	660	440	100	18,18	1,83
5	120	720	480	100	19,44	1,71
6	130	780	520	100	20,51	1,63

As a result of expanding the product range, production and sales volumes increase, which in turn leads to an increase in the operating profit of the organization (figure 1.4).

The dynamics of the *DOL* ratio indicates an increase in the operating profit of an organization with a change in revenue by 1%. For example, *DOL* = 2 (at the third stage) means that a change in revenue, for example, by 5% leads to a change in operating profit by 10% (5% • 2). The operating leverage is also directly related to the level of operational

---

6 Ibid.

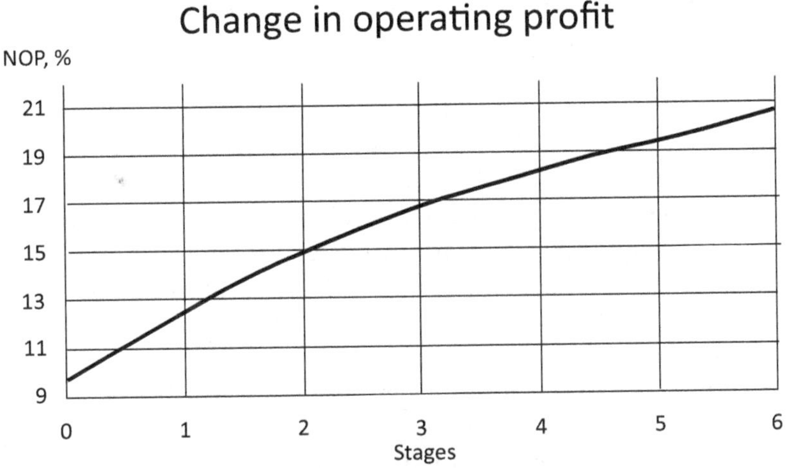

Figure: 1.4. Change in operating profit due to expansion of the product range

risk: the greater the operating leverage, the greater the risk. In this regard, a decrease in the *DOL* ratio from stage to stage also indicates an increase in the economic stability of an organization (figure 1.5).

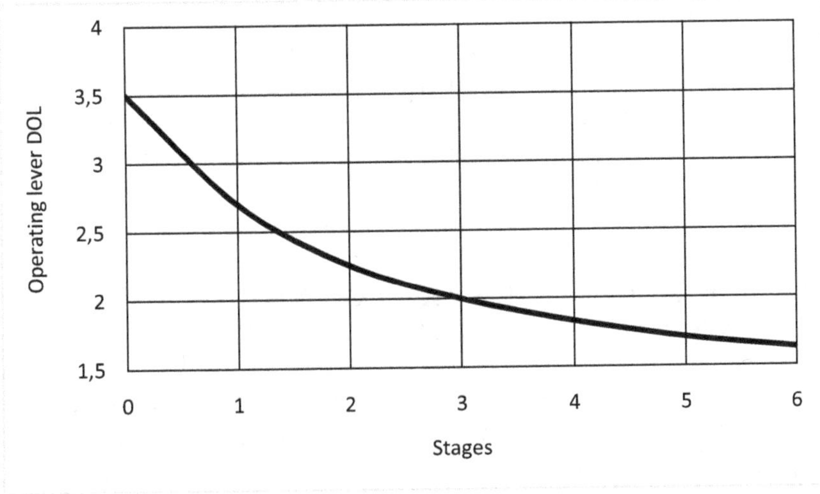

Figure 1.5. Gradual decrease of the *DOL* ratio, indicating an increase in the economic stability of an organization

Thus, the increase in the efficiency of an organization's functioning illustrated by us, in turn, indicates an increase in the level of its economic stability as a result of the expansion of the product line.

This process is defined by the main parameters of an organization's competitiveness, which determine the creation of products that meet consumer expectations. Based on the necessary costs for the development and transfer of competencies, development of innovative technologies and modernization of production, the competitiveness of products in the market is determined as a measure of their purchasing power, given the timing of product launch. In this sense, product competitiveness is a complex synthetic concept that reflects the efficiency of organizational and economic processes of development, pre-production, production, sales, after-sales service, etc. Therefore, methods for achieving high product competitiveness must take into account the organizational and economic aspects of its creation, various risks, and manage all processes from shaping the image of a product to the moment it is replaced with a product with a higher consumer value with better consumer properties.

Thus, sustainable economic development of an organization is based on timely adaptation to environmental conditions by introducing the listed technologies into production in order to increase labor productivity, production efficiency and ensure the competitiveness of products.

1.2. SHAPING THE IMAGE OF NEW PRODUCTS BASED ON MARKETING INFORMATION

Considering the issues of sustainable economic development of an organization, we can conclude that it should update products depending on the requirements of the market and consumer expectations. In this regard, it is necessary to have a clear understanding of the dynamics of the main market indicators and new requirements for products from the consumer, the characteristics of competing products, as well as new technical solutions, know-how and key competencies of competing companies. As part of our research, we will consider the advanced development of an organization as a "leap" made due to the accumulated intellectual potential and competencies implemented in fundamentally new technical, technological and software solutions; and the process of managing the advanced development of an organization in an already created (occupied) market of high-tech products – as a purposeful collection, systematization and analysis of specific information about the internal potential of an organization and the external environment in order to find and rationalize the existing competitive advantages and minimize risk flows for effective management of product and product line updates. These issues are extremely important due to such processes as rapidly growing competition in the global high-tech market, loss of market share and changes in consumer preferences, which can negatively affect the economic stability of an organization.

Let us state the main task of product renewal management based on the characteristics of the life cycle of a knowledge-based product. One of the main problems of managing the advanced development of a high-tech organization is the significant time gap between the fundamental scientific research, the manufacture of a prototype of a high-tech innovative product and the starts of its mass production. Already at the stage of research, it is necessary to set a trajectory for improving the techno-economic parameters of products based on updating. In the conditions of dynamically changing and highly competitive high-tech markets, analysis of the growing needs of society and emerging new competencies makes it possible to predict the development of the market with an assessment of the prospects for updating products. In this regard, a deep analysis of the promising needs of organizations and society is required both at the earliest stages of the life cycle of high-tech products, and in the process of product release and preparing for its modernization.

Marketing research of new needs of organizations and society is one of the management tools that help reduce risks under conditions of uncertainty. By demand forecasting, an organization can shape the image of a promising competitive product and determine its techno-economic characteristics (figure 1.6).

Features of the organization of effective marketing research aimed at ensuring the competitiveness of the developed products are:

1. stage of forming a methodological base for research. In the generalized structure of marketing analysis (Fig. 1.6), the methodological base is formed depending on the stage at which the research is conducted. The entire methodological complex can be divided into 2 groups corresponding to different stages of the study:

   • marketing research methods defined in the plan as a way to achieve the goal. At the same time, the choice of the

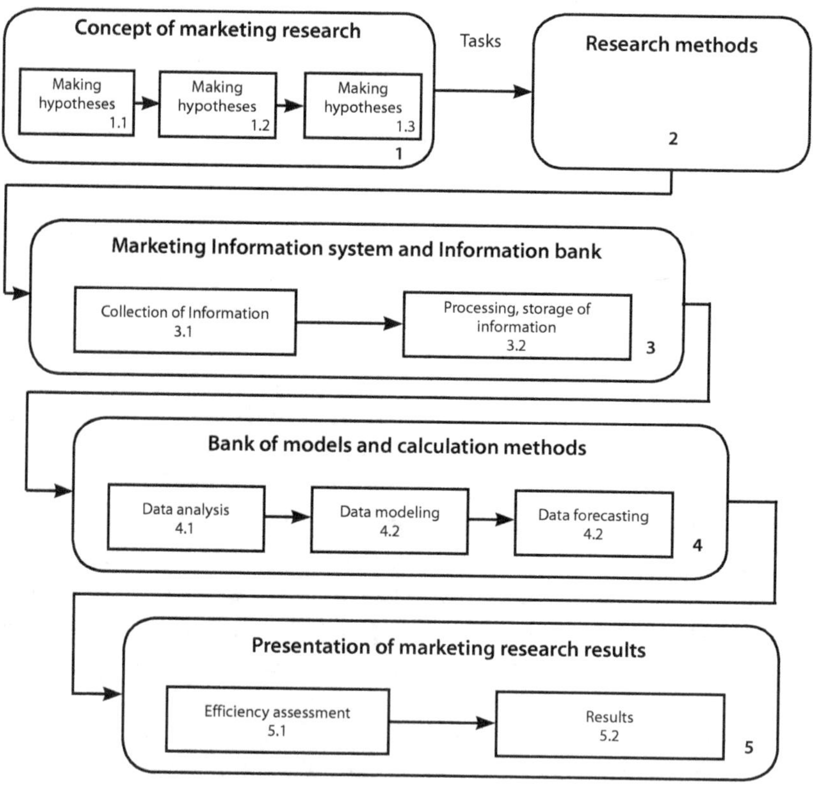

Figure 1.6. Generalized scheme of marketing research

methodology to be used for marketing analysis is determined primarily by the problems formulated at the preliminary stage and the overall goal of marketing research;

- methods used for systematization, analysis and subsequent assessment of the collected marketing information.

According to fig. 1.6 the choice of marketing research methodology is the result of the development of a holistic integrated research plan. Along with the definition of the type of necessary information, the most representative sample size for solving problems and methods of collecting the necessary information, it precedes the stage of implementation of the actual marketing research as such. In this scheme, the choice of all methods is the competence of the performer and is determined by the goals and problems of the research, which makes it possible to more clearly define the task and ways to solve it, as well as adequately assess and plan the cost of research, which is important if research is conducted frequently;

2. stage of tracking the results, i.e. introduction of the information received into the overall management system of an organization. Often, the qualitative information obtained as a result of the research turns out to be useless, since the leaders of organizations do not have experience in the effective use of the presented results in the management process.

At the present stage of development of existing approaches to the management of economic objects and taking into account the existing scientific and technological reserve in the branches of high-tech production, one of the effective mechanisms for improving competitiveness can be the introduction of an integrated management system for updating the product range. It is based on the economic laws of the functioning of market relations and general scientific methods of management. To ensure the competitiveness of high-tech products in the world market, it is necessary to improve the quality of strategic marketing carried out by an organization within the general concept of advanced development management in the context of product renewal. Ensuring competitiveness in the global and domestic markets is carried out through continuous modernization and management based on the introduction of innovative technologies in the production process and its management using complete and objective initial information about the current market conditions.

The effective use of the comparative advantages by organizations is always directed towards those areas that favorably distinguish their products from those of potential or real competitors. Marketing research forms a complex of necessary information about the external environment and internal capabilities of an organization.

The schematic diagram of any modern automated control system based on analytical processing of a wide range of input data about the external environment, capable of assessing risks proves the importance of building such a complex of specialized information about the needs of society and organizations, markets, and the potential of a manufacturer. Since any automated system always operates based on an algorithm created through a mathematical description, it is advisable to use system input data, which in this case represent the quantitative and qualitative results of targeted marketing research in the software complex for imitating economic and mathematical modeling of the advanced development management process.

For successful high-tech industrial organizations, all marketing information is collected, analyzed and distributed within the marketing information system (MIS), which is part of the overall information system for managing an organization's advanced development. Thus, a marketing information system is a set of staff, equipment, procedures and methods designed to collect, process, analyze and distribute timely and reliable information necessary for preparing and making management decisions[7]. To increase the competitiveness of a high-tech product, a manufacturing organization needs a special MIS structure. The crucial links between managed objects and the direction of information flows and management impacts in the proposed marketing information system focused on managing the advanced development of the organization are shown in figure 1.7.

---

7 Chursin A.A., Okatiev N.A. Innovation and investment in the organization M.: Engineering, 2010.

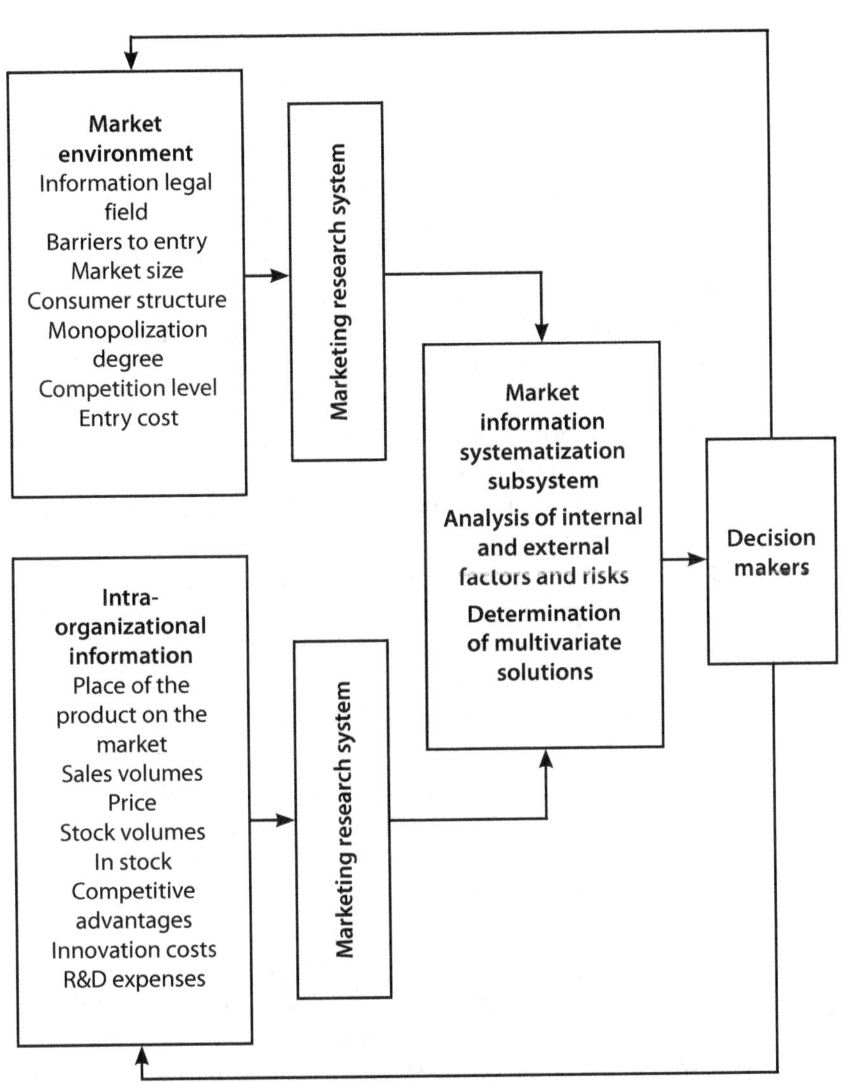

Figure: 1.7. Marketing Information System (MIS)

In order to structure the process of achieving the competitive-ness of high-tech organizations, we highlight a number of features of the world market for high-tech products, systematized on the basis of the analysis of marketing information.

25

1. The market for high-tech products is formed by high-tech goods produced in the fifth and sixth waves of innovation. Currently, the sixth wave of innovation is being widely industrialized, the core of which includes nanoelectronics, genetic engineering, multimedia interactive information systems, high-temperature superconductivity, fine chemistry, etc. High-tech products form almost the entire export volume of industrialized countries.

2. Imports of high-tech products in most industrialized countries are limited by serious barriers.

3. Prospects for the development of the domestic market for high-tech products lie in the area of the necessary expansion of the volume of its export, provided, among other things, by elements of price competition, to traditional and currently opening world markets.

4. Due to the monopoly on intellectual property, manufacturers of high-tech products may often have no direct competitors.

5. The global market of high-tech products is formed by industries with the highest rates of advanced development, acting as carriers of advanced innovative technologies, the latest methods of organizing production and management, which are closely related to the development of relevant scientific areas and change depending on the degree of maturity of the basic technologies and the phase of the industry development cycle.

6. Most of the high-tech industries on the world market are simultaneously related to the industries that ensure the defense capability and technological independence of the producing country, i.e. the state of the market situation in the world market directly affects the national security of the producing countries.

7. Many types of high-tech products on the market are characterized by a long life cycle, reaching 10-15 years or more.

The process of managing product updates should be organized based on the above-mentioned features of the high-tech products market. Next, we will form the key components of the product update management process as a step-by-step diagram (figure 1.8).

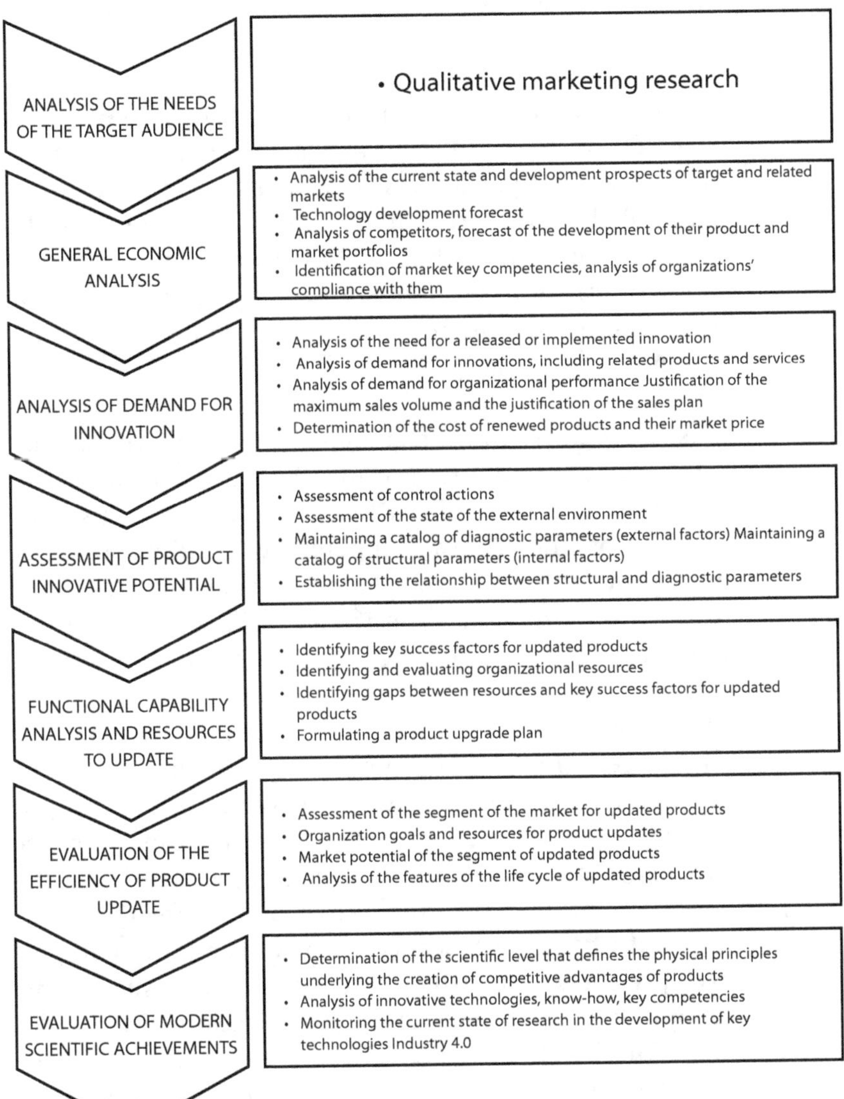

Figure: 1.8. Generalized diagram of the product update management process

Let us highlight the main features of the process of studying prospective needs in the context of updating the products of high-tech organizations. Updating products may involve adapting to new markets (for example, foreign markets).

1. To organize an effective product update process, it is necessary to have sufficiently accurate marketing information about the markets in which the updated products will be presented. It is necessary to forecast the dynamics of market changes based on identifying indicators of transformation of consumer preferences, taking into account the requirements for product personalization.

2. The duration of the life cycle of most high-tech products, as well as the effect of various geopolitical factors and sanctions, determine the need to target not only the markets of high-tech products, but also the markets of know-how, competencies, leading achievements of scientific and technological progress, the world's fundamental scientific groundwork in the relevant areas, i.e. marketing research of the innovation market.

3. The analysis of the market and consumer expectations allows determining the image of promising products. Further actions of a company to achieve a dominant position in the market are associated with creating resource support, the development of key competencies and the development of innovative technologies. Effective management of these processes will allow creating competitive products faster than competitors. The sustainable economic development of an organization directly depends on the effective use of innovative technologies for the release of existing and new products, as well as on the methods of creating competitive advantages of products and services.

1.3. INTELLIGENT MONITORING OF THE EXTERNAL AND INTERNAL ENVIRONMENT AS THE BASIS FOR CREATING THE IMAGE OF NEW COMPETITIVE PRODUCTS ON THE MARKET

Currently, the problem of increasing the competitiveness of high-tech industries through the active introduction of digital technologies and artificial intelligence in the management cycle and at all stages of the product life cycle is widely studied. In this context, it is important to substantiate the mechanism for integrating the latest online monitoring platforms into the system for launching new products.

Intelligent monitoring (IM) actually ensures effective performance, generation of innovations and competitive positions of an organization, thereby guaranteeing its economic survival and market success.

An intelligent monitoring system can be considered as a set of the following subsystems:

1. collection and storage of parameters from various automated systems of the entire production complex;
2. a database consisting of information and dynamic data;
3. monitoring of changes in the state of equipment, complex, production as a whole;
4. calculation and analytical work;
5. technological document flow;
6. information protection;
7. user interface.

The most important task of developing high-tech industries (HTI) is to increase their competitive advantages based on:

- influence of external and internal factors on the competitiveness of the designed products, which would avoid design errors and thereby reduce unnecessary costs;
- setting more advanced technical parameters of the designed products, new requirements for technological processes and the level of production development;

- improving quality parameters earlier and more accurately than competitors, providing a certain competitive advantage and advanced market entry.

Practice thus shows that an increase in investment in quality improvement at the design stage by 2% gives an increase in profit from the sale of finished products by 20%. Therefore, in order to increase the competitiveness of high-tech industries, it is important to conduct monitoring based on digital and intelligent technologies.

Monitoring is a means of information and analytical support of management. The development of the knowledge-based economy has expanded the functional field of monitoring in the form of a mechanism for generating the latest knowledge about objects and management processes.

Intelligent monitoring as a type of management activity is designed to solve the following tasks:

- creating additional value for information through the definition, selection, identification, storage and dissemination of knowledge;
- endowing knowledge with consumer properties in order to provide them in a form accessible to the user;
- use of knowledge in manufacturing, services, preparation of documentation, databases;
- motivation for building knowledge (internal dissemination, interchange between employees);
- creating an interactive training field among employees for regular exchange of information and using all conditions for building knowledge;
- knowledge assessment and operation of an organization's intangible assets.

Today, many successful organizations have incorporated intelligent monitoring into their strategic planning and management structure.

The use of end-to-end intelligent technologies in monitoring expands its capabilities to ensure the competitiveness of high-tech products and the manufacturer at the design stage, but it is accompanied by a number of challenges. One of the most critical challenges is the complexity of introducing new digital platforms into the existing information infrastructure. In this regard, the model of transferring an organization's IT infrastructure to a new platform (combined with intelligent monitoring) with a built-in integration mechanism, as well as a transition scheme, is used. The practical application of the developments enables moving to a new level of competitiveness with high economic feasibility of high-tech products.

In the context of digital transformation and the use of artificial intelligence, there are opportunities for implementing a fundamentally new approach to ensuring the competitiveness of high-tech products at various stages of the product life cycle, including at the stage of preliminary design. The potential of digital technologies can be realized if the life cycle management system of high-tech products is transformed, primarily by creating effective mechanisms to ensure the development and production of competitive products.

Development of products in virtual mode, development and operation of a product, in other words, preliminary modeling of production processes and the formation of virtual structures, allow a quantitative preventive assessment of the competitiveness of products.

The main purpose of intelligent monitoring is to receive, store and process information from the external and internal environment, to regulate and coordinate the activities of the entire set of elements of production systems. During work, the functions of intelligent monitoring were determined that ensure the sustainable competitiveness of a high-tech organization:

- integrative – combining various components of the production system (PS) into an integrated unit and ensuring their interrelated and coordinated functioning;
- coordinating – maintaining the constancy of the internal environment, ensuring the coordinated flow of functions of various elements of the PS, adapting them to changing environmental conditions;

- regulatory – ensuring the ability to restore and maintain internal stability as a result of reactions that compensate the influence of external influences;
- information and communication — the implementation of regulation of the intensity of exchange in a kind of high-speed highways that transmit a huge amount of information every second. Information is generated in various sensors and special centers of origin and processing, then it enters the control center, where ideas about the external and internal state of the production system are generated;
- adaptation – restructuring the activity of various elements of the production system in accordance with changes in the environment based on the analysis and synthesis of various information coming from specialized sensors;
- prospective – forecasting the future behavior of objects and subjects based on current and retrospective facts, identifying risks and opportunities for predicting future events.

The realization of IM functions is based on the integration of complexes of unified information models and the use of specialized software.

Intelligent monitoring is a base element of the information and communication system of knowledge-based and high-tech organizations, which includes: diagnostic complexes, databases (DB) and knowledge bases (KB), expert systems (ES), modeling systems, business intelligence and visualization platforms.

The digital platforms provide not only analytical processing of a huge data set, but also building forecasts and models of the future behavior of the production system and its individual elements necessary for making informed management decisions (MD) in real time.

The main tasks of the module *"Monitoring the external environment"* include:

- assessment of potential demand for high-tech products;
- requirements of potential customers for high-tech products (new competencies and characteristics);

- analysis of competitors and their products;
- assessment of a company's competitive position in the market.

As stated by *Baimuratov, Makarov & Chursin, 2011*, assessing the competitiveness of high-tech products is a necessary condition in the process of an organization's work to bring it to market. Subsequently, information is generated for designers in terms of competitive techno-economic characteristics (parameters) of the future model of high-tech products (Fig. 1.9).

The "R&D Monitoring" module, in terms of determining the competitiveness parameters of the future model of high-tech products, should primarily perform tasks related to:

- tracking information flows about trends in the technical and technological development of high-tech products, identifying problems that require urgent solutions, and searching for possible solutions to determine the trajectories of developing processes in various fields of knowledge;
- support of complex products at all stages of the life cycle, up to disposal;
- assessment of the current state of the problem based on a comparison of information received from the external environment and assessment of the capabilities of a particular organization to create competitive high-tech products, *(Ryzhov, 2008)*;
- forecasting, modeling technical development and trends in the competitiveness of high-tech products, forming on this basis the techno-economic image of a new model;
- planning and control of R&D execution;
- searching for ways to reduce the labor intensity of R&D and, accordingly, reduce their cost, which will significantly affect the speed of launching new products to the market, in particular, the use of digital and additive technologies, as well as interactive design methods.

The *"Technological monitoring"* module is designed to timely identify opportunities for technological development and achieve competitive advantages based on:

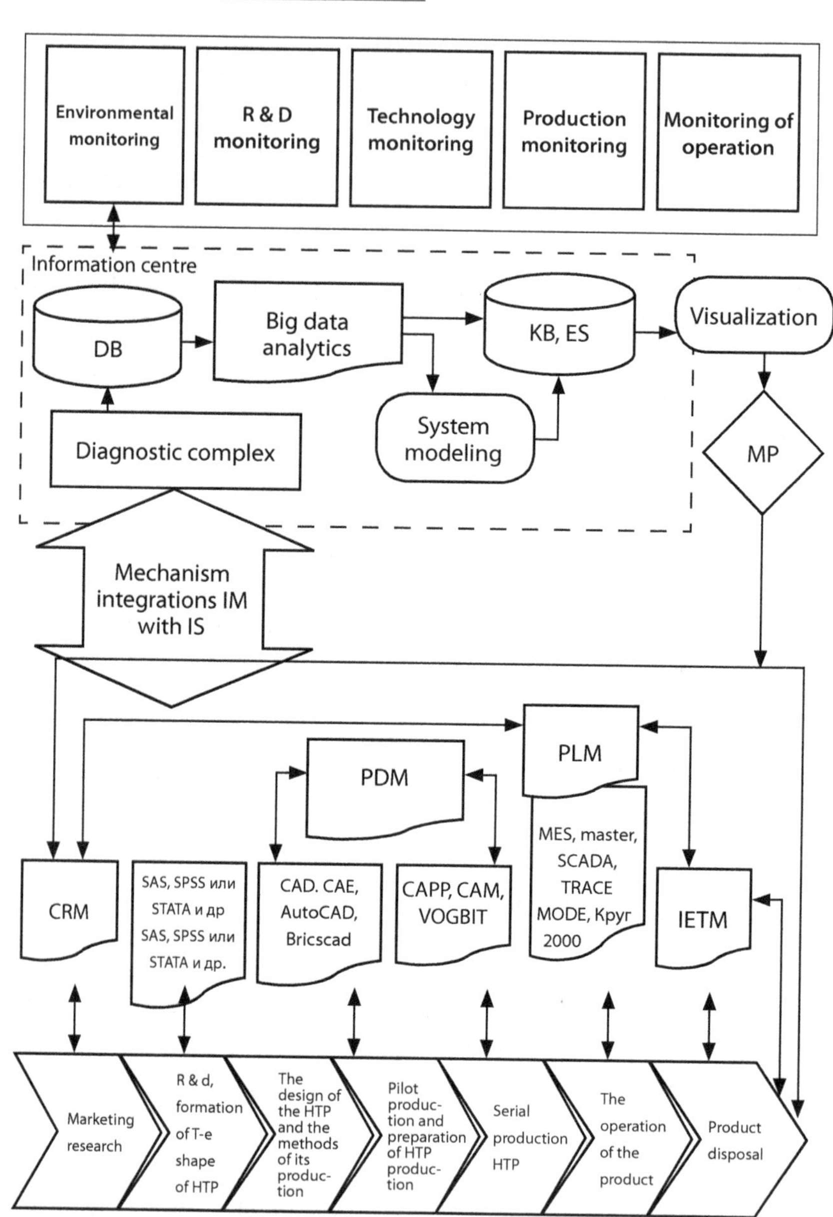

Figure: 1.9. Model of an organization's IT infrastructure for a new platform (combined with intelligent monitoring)

- comparative analysis of own technological equipment with the equipment of competitors;
- tracking development trends and opportunities for new technologies capable to provide a breakthrough in competitiveness;
- obtaining and accumulating the initial information necessary for developing strategies for technological development and simulation of their their implementation;
- development of recommendations to reduce external and internal risks when introducing new technologies[8].

The intelligent module *"Production monitoring"* is possible if all production facilities – equipment and workplaces – are combined into a single information network. This ensures the possibility of using the IoT – the industrial Internet of things. It is enough to install sensors with a network connection, and conditions are provided for manufacturers and customers to remotely monitor the operation of production sites, timely carry out routine maintenance, predict accidents and perform scheduled preventive maintenance, etc. Thus, the entire chain from suppliers to consumers will be carried out without human intervention. *"Equipment Monitoring"* is *a* variation of this intelligent module, which performs the following functions:

- equipment loading control;
- fixation and typology of equipment downtime;
- inspection of labor discipline of personnel;
- warning of emergencies requiring a quick response by e-mail messages or notifications from managers on the work computer;
- dynamic analytics for a quick assessment of the situation, key performance indicators (KPI) of an organization, its departments, individual pieces of equipment and operators;
- visualization of analytics results and generation of reports with a visual display of time losses and reasons for equipment downtime, etc.

---

8 Efimushkin I.V. Expert information system of technological monitoring //T-Comm – Telecommunications and Transport. – 2009

As a result of performing these functions, "bottlenecks" are promptly identified, time losses and their causes are assessed, equipment load is optimized, working time reserves are created, etc. The module can be combined or integrated into the production monitoring module.

The module *"Monitoring the operation of high-tech products"* is becoming crucial in increasing the competitiveness of products and of a manufacturer, since using the methods of Data Mining, an organization not only takes responsibility for the efficiency of operation, but also ensures a decrease in operating costs and reliability work due to continuous monitoring and diagnostics of the technical condition of the object. Timely detection of defects in the functioning of the object and their elimination becomes a decisive factor in improving the efficiency of operation and competitiveness of an organization.

Integration of intelligent monitoring with existing information systems in high-tech production is not actually adverse. On the contrary, modern monitoring systems of the MDC/MDA (Machine Data Collection / Machine Data Acquisition) class make it possible to improve production without significant investments. The study developed a scheme for the transition to a higher-quality level of competitiveness of high-tech products based on intelligent monitoring (Fig. 1.10).

The tasks are solved not only to increase the efficiency of production, but also to address a number of related problems. The most famous and large-scale global platform solutions are developed by Dassault Systèmes and Siemens PLM Software. Siemens Teamcenter software includes a platform and portfolio of applications focused on a specific range of tasks. The platform is the core of the system with a set of basic components. It is a system "bus" for deploying various applications. Teamcenter allows the management of various types of data describing a product at all stages of its life cycle: 2D and 3D data created in various design systems (including CAD systems for microelectronics), graphic and text documents in any format, sets of attributes combined into electronic cards, and other types of information presentation. Teamcenter software operates on a single product and process database. At the same time, the joint teamwork

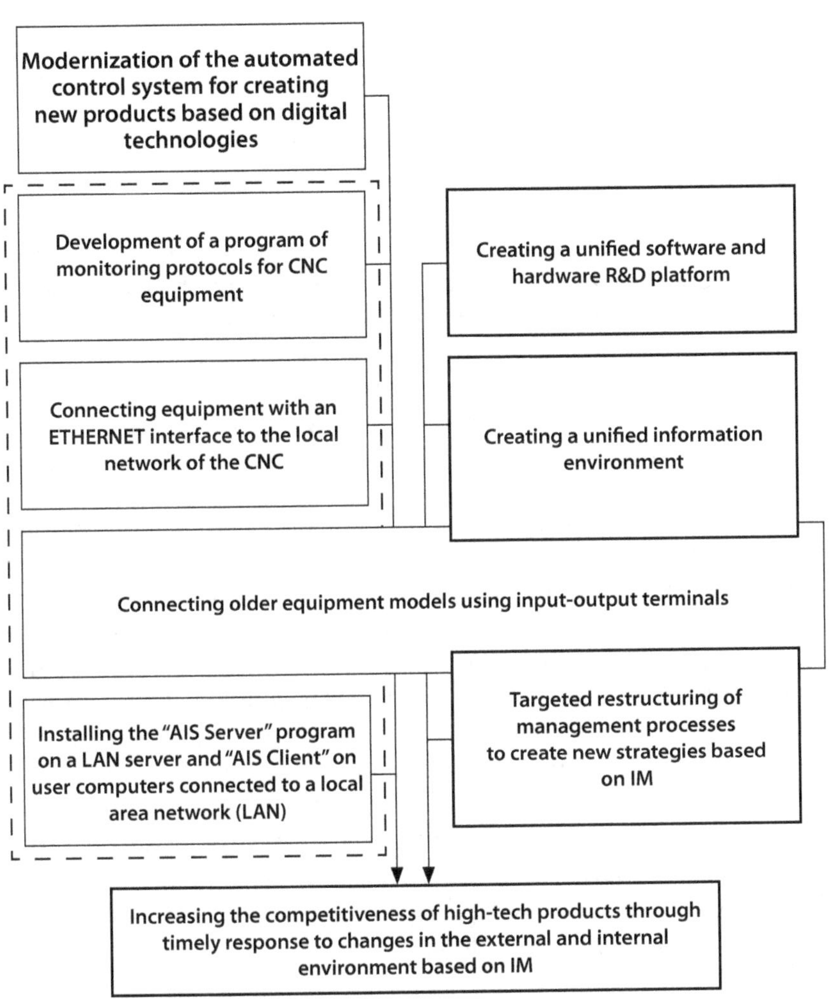

Figure: 1.10. Scheme of transition to integrated intelligent monitoring (IM) in a unified information environment of high-tech products

of designers on a single electronic product model and parallel work on the product of design and technological services is provided.

Both Siemens Teamcenter and Dassault Systèmes 3DEXPE-RIENCE global platforms provide role-based data access, cloud-based platform operation, and scalability. The integration of such systems with ERP, PDM, EAM, etc. consists in the synchronization of current changes in parameters to guarantee the relevance of reference and other operational information, eliminating errors and ambiguity of generated reports. The analysis of the compatibility of software products showed the achievability of the proposed developments and their practical significance.

The proposed concept of cost-effective intellectualization of monitoring allowed, through the analysis of the transformation of its functions and the compatibility of software products on the market, development of a mechanism for integrating an information system with intelligent monitoring, built into the model of transferring to new platforms, as well as a step-by-step algorithm (diagram) of the transition to IM in a single information environment of high-tech products.

The blocks of integrated intelligent monitoring presented by stages of the life cycle ensure timely and preventive receipt of objective, reliable, relevant, predictive and new-knowledge creating information necessary for making management decisions that directly affect the competitiveness of high-tech enterprises. In the presented IM system, control over the equipment and the production process is performed automatically according to the incoming parameters of changing the state of an object. Input information is compiled in the form of a regular data collection cycle due to the fact that control actions on production may be provided from none to partially. Therefore, the main task of intelligent monitoring is to determine the discrepancy between the planned and actual parameters of management actions[9].

---

9 Efimushkin I.V. Expert information system of technological monitoring // T-Comm – Telecommunications and Transport. – 2009.

CHAPTER 2. MANAGING THE DEVELOPMENT OF PROMISING HIGH-TECH PRODUCTS

2.1. BASIC PRINCIPLES OF NEW PRODUCT DEVELOPMENT MANAGEMENT

The use of modern methodological approaches aimed at solving the problem of designing products for a given cost and competitiveness in the market is the basis for the production of new highly competitive products. The use of such software solutions provides competitive leadership to high-tech organizations in developed countries by building effective production activities with optimal allocation of resources (material, human, intellectual competencies, etc.) at all stages of the life cycle of a project for the development, production and sale of high-tech products during given technical requirements and economic parameters.

A feature of the process of creating innovative products that claims to form a new market, along with the development of a technical project is the extensive preparatory work for the technical re-equipment of production, the development of new equipment in order to drastically reduce the development time and pre-serial production.

The reference point in the organization and conduct of such works is the achievement of indicators close to those provided in the design technology for products (labor intensity, material consumption, etc.) by the start of serial production, due to rational interaction with design bureaus that develop products, organizations-manufacturers of products, central and sectoral technological structural units at all stages of work (Fig. 2.1).

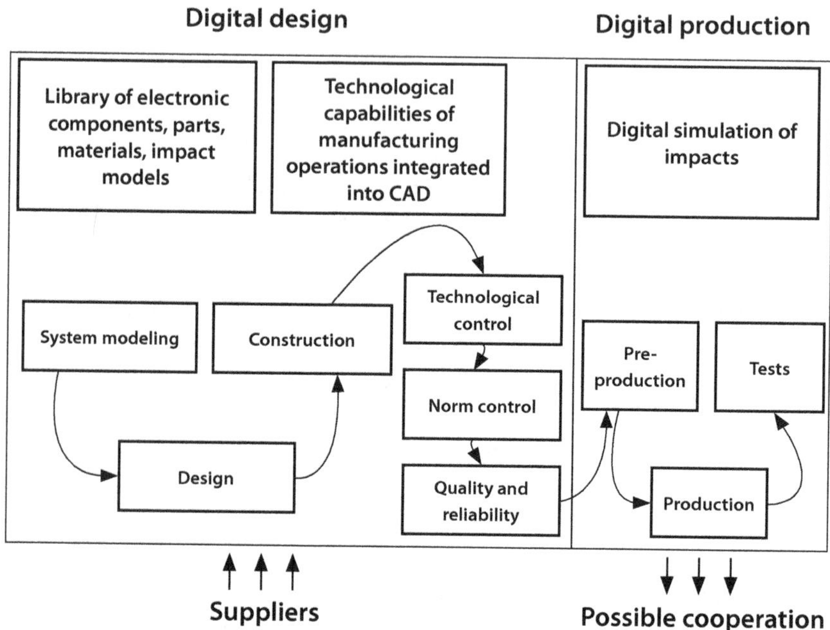

Figure 2.1. End-to-end digital design and production

The principle of parallel (combined) work within a single end-to-end program of the cycle "advanced development products – production – product renewal" ensures a high level of technology development on an improved product (with minor deviations in the initial labor intensity) as soon as possible after the start of serial production[10].

In terms of the impact of digitalization on the processes of creating and manufacturing products, there are trends in reducing the life cycle. For example, the use of additive technologies can reduce the stage of pre-production, and ensure an earlier release of the product to the market. The released time lag can be used by an enterprise to implement advanced development measures. These trends indicate that the

---

[10] Razumovsky V.A., Okatiev N.A., Chursin A.A. Approaches to optimizing resources for the creation and production of competitive products // Handbook. Engineering journal with annex. – 2008. – No. 5 (134). – Pp. 35-40.

deepening of informatization processes in all spheres of life inevitably entails the development of technologies and the needs of society, and therefore the priority directions in the development of science, technology and technology are changing in the world. In order for such priority areas to be promising and their choice to be carried out with regard to the latest economic trends, it is advisable to implement the mechanism for forming priority areas presented in Figure 2.2.

In practice, the implementation of this mechanism using Data Mining methods will make it possible to create a list of the most promising technologies that meet the requirements of the market,

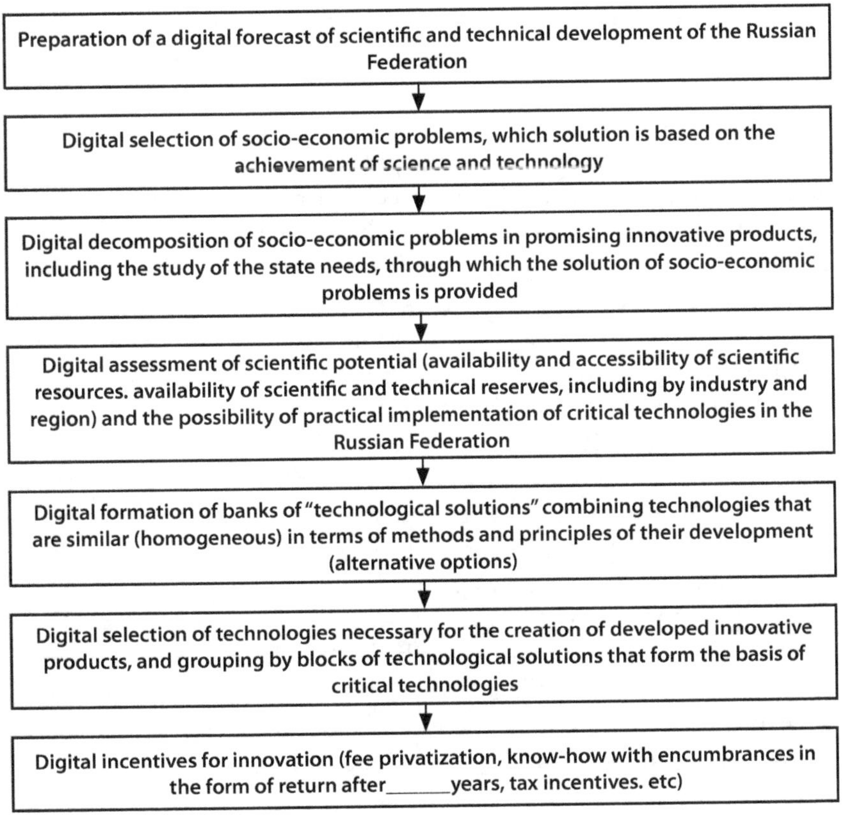

Preparation of a digital forecast of scientific and technical development of the Russian Federation

↓

Digital selection of socio-economic problems, which solution is based on the achievement of science and technology

↓

Digital decomposition of socio-economic problems in promising innovative products, including the study of the state needs, through which the solution of socio-economic problems is provided

↓

Digital assessment of scientific potential (availability and accessibility of scientific resources. availability of scientific and technical reserves, including by industry and region) and the possibility of practical implementation of critical technologies in the Russian Federation

↓

Digital formation of banks of "technological solutions" combining technologies that are similar (homogeneous) in terms of methods and principles of their development (alternative options)

↓

Digital selection of technologies necessary for the creation of developed innovative products, and grouping by blocks of technological solutions that form the basis of critical technologies

↓

Digital incentives for innovation (fee privatization, know-how with encumbrances in the form of return after_______years, tax incentives. etc)

Figure: 2.2. Mechanism for the development of priority areas of science, technology and technology capable of satisfying personalized needs, thereby stimulating organizations to develop promising products of the "future", to build new market niches.

The basis for the innovative development of a modern organization is the timely introduction of innovations in the design of advanced development products.

In parallel with innovative development, the production is increased and the required output of finished products is ensured by eliminating "bottlenecks" and imbalances based on the introduction of new technology. At the same time, the new technology should have sufficient competitive advantages to establish prerequisites for the development of new samples in the future, as well as interface with the design technology of the samples being developed, thus ensuring a technological balance.

Measures for technical re-equipment, reconstruction and expansion of production, being tied to the production structure, to the terms and order of execution, should be checked for completeness and availability of resources, for compliance with the requirements for production using a system of balance equations (balances of labor resources, capacities, production areas, technological equipment). The system that meets these requirements and is refined during its implementation to reflect changes in the goals and conditions of development of the organization is a program of continuous development of its material base.

This program initially (at the stage of development) forms part of a comprehensive target program for the development of an organization, covering all aspects of social and economic development, but during its implementation it is allocated into an independent document that has certain specificities.

The mechanisms of innovative development, together with the modernization of production facilities, lay the resource, informational and organizational basis to ensure the advanced development of technologies, the concentration of resources for the consistent removal of bottlenecks and the elimination of imbalances in industry production.

An organization should carry out its actions on the basis of the creation of resources and innovative potential, which will lead to sufficient profit for the modernization of production, the development of new products, and stable economic development.

To make objective decisions, it is necessary to take into account numerous factors that affect the techno-economic activities of an organization, including factors related to the adoption of a particular model of economic calculation. For this reason, it is necessary to develop a mechanism for economic management of the processes of technical re-equipment of an organization for the production of new types of products corresponding to the world technical level.

Currently, these actions are carried out only on the basis of a comprehensive analysis of techno-economic activities and cost optimization for the resulting revenue effect. This approach allows having a stable economy and the necessary development, meeting the customer needs and requirements;

In modern market conditions, when releasing new types of products, it is particularly important to consider such indicators as capital productivity, capital-labor ratio and profitability, which largely depend on the results of the introduction of R&D, new technology, technical re-equipment and have a significant impact on many forms and economic indicators of organizations and overall production efficiency. Moreover, it should be noted that the forms of economic implementation of efficiency, for example, fixed productive assets, are versatile: it is an increase in labor productivity, a decrease in material consumption, and a reduction in production costs. Improving the efficiency of fixed assets also affects the revenue growth, reducing the cost of production. As capital productivity is a function of capital-labor ratio and labor productivity, the relationship between the movement of the value of a unit of production capacity and capital productivity is of an indirect nature. However, it must be recognized that in many organizations the indicators of the use of fixed assets are not standardized and, therefore, the accounting of the need for fixed assets is not always accurate. Thus, the impact of the use of fixed assets on the change in cost can be calculated through generalizing a indicator – capital productivity. The total increase in output is set as follows:

$$\Delta\Pi = (C_{\varphi2} - C_{\varphi1})\, f_1 + (f_2 - f_1)\, C_{\varphi2},$$

where $C_{\varphi1}$ and $C_{\varphi2}$ are the average annual cost of fixed assets in the base and reporting periods; f_1 and f_2 are capital return in the base and reporting periods.

When calculating according to this formula, a change in output due to a deviation in capital productivity is possible as a result of an improvement or deterioration in the efficiency of fixed assets. At the same time, in the practice of economic calculations there is a methodological technique which is carried out through a change in the conditional-fixed part of the overhead cost when the determination of the economic effect in the case of the introduction of technical solutions aimed at increasing the volume of production. Knowing the total amount of the conditional and fixed costs $C_{c.f}$, you can calculate their level (expenditure) $E_{c.f}$ per ruble of production output (A_{p1})

$$E_{c.f1} = C_{c.f1} / A_{p1}$$

Then the change in the cost as a result of the improvement (deterioration) in the use of fixed assets is

$$\Delta C = \left(f_2 - f_1\right) C_{\varphi2} \frac{C_{c.f1}}{A_{p1}}.$$

Replacing the difference for Δf, we get

$$\Delta C = \Delta f\, C_{\varphi2}\, E_{c.f1}.$$

When planning economic efficiency and other indicators of the use of new technology in production, it is advisable to use the profitability indicator. In its economic essence, the economic effect from the results of R&D, the introduction of new equipment and technologies is a practical increase in profit, and the total costs of their development and development is an increase in fixed assets. Thus, the implementation of R&D results can be expressed in the growth of the profitability indicator compared to the established average for the industry.

The level of profitability directly or indirectly depends on the entire set of production and financial factors: the level of labor productivity, capital productivity, the speed of turnover of working capital, etc. Through the growth of profitability as a result of the introduction of scientific and technical measures, the economic efficiency of innovative development will also be associated with the existing economic mechanism for stimulating production.

At the same time, the need for an integrated approach to innovative development does not negate certain features in its various links. One of these links is the output of new progressive products that are highly competitive on the world and Russian markets.

The new developed product should have breakthrough competitive advantages: a more perfect principle of action, purpose, improved quality, a set of improved properties or significant changes in the most important indicators.

In practice it is apparent that the risks of introducing new technologies when launching products are either the result of existing internal conditions of the production itself, or the result of factors that lie outside of technological progress, such as errors and shortcomings in pricing. However, possible actions of competitors to develop similar products should be taken into account. It is necessary to monitor such processes and react to them promptly, including by adjusting the technical characteristics of the products being developed.

Further, we will consider the issues of how an organization achieves a level of innovative potential sufficient for the production of highly competitive products. Herewith, the innovative technologies formed during the creation of new products should be based on a certain set of knowledge, which is transformed into the competence of organizations and their teams. In reality, some components of the innovative potential may reach a high level, while others may stay at a lower level. The lack of innovative potential is eliminated through the development of new competencies. At the same time, the effectiveness of the built competencies and innovative technologies developed on their basis is assessed in terms of the technical and cost characteristics of products achieved through their use.

2.2. DEVELOPMENT OF KEY TECHNOLOGICAL COMPETENCIES TO ENSURE THE CREATION OF PROMISING COMPETITIVE PRODUCTS

It is generally accepted that industry 4.0 requires the design of new strategic approaches to the introduction of technologies, building competencies necessary for the development, production and launch of products to the market. The development of unique competencies provides a significant synergistic effect in the economy, which is manifested not only in industry, but also in education, as the demand for competencies and a high level of education also generate offers for new educational services. Moreover the development of competencies and the emergence of new technologies create conditions for new consumer markets. In contrast to traditional industry competition in the field of high technologies and unique competencies has other mechanisms that react much faster to changes in competitors' activities, which significantly motivates companies to quickly and effectively change their behavior.

The efficient functioning of knowledge-based organizations creates demand for innovative products. Consequently, there is also a demand for key competencies for new industries. Of course, the increase in this demand contributes to an increase in the supply of key competencies, the development of fundamental and applied science, generates applied developments that form the basis of innovative technologies, which are then used to create new products or services. The drive to commercialize innovative technologies forces high-tech companies to create or propel new markets to promote their new products. As a result of the creation of new consumer markets for the sale of innovative products, the rapid growth of relevant industries begins, a large number of relevant key competencies are built, which in turn lead to the emergence of more advanced products and, ultimately, new consumer markets.

The generally accepted concept of competence is the result of identifying the knowledge, skills, competencies and experience of employees of an organization in a complex. But to create promising competitive products in high-tech production, key technological competencies are important. Thus, one of the main directions of an organization's strategy should be the creation and strengthening of such competencies, as a result of which it will be possible to develop competitive products.

Conditionally, three levels for maintaining an organization's leadership in the competition can be identified: competition in the market of finished products; competition of key products; competition of key technological competencies (Figure 2.3).

Key products are actually the physical embodiment of key competence, which undoubtedly provides an organization with its long-term competitiveness in the market.

In the context of high financial costs for the development of new technologies, the issue of effective use of all types of resources should be given special attention. Based on the methods of assessing key technological competencies and selecting innovative technologies, it is possible to build a research program and determine the most relevant areas of financing the development of new competencies and technologies. To do this, the mechanisms for managing the development of competencies should provide for their replication to create innovative technologies in various areas

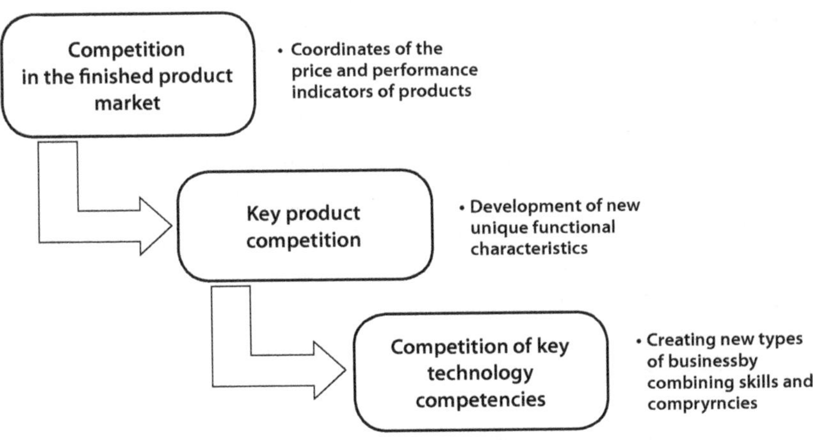

Figure: 2.3. Levels of maintaining an organization's leadership
in the competition

of development. For example, in the helicopter industry, the successful application of a technology in a particular area does not guarantee its similar successful implementation in another area. In the context of economic efficiency, the processes of competence transfer and the organization of specialized competence centers that combine production resources and relevant competencies for their operation are of particular importance. Such centers allow creating new projects from individual elements, which can reduce the cost of resources for parallel divisions. The advantage of this activity is that such actions can be used both within one organization and by involving temporary structures, for example virtual organizations. In this case, major specialists can be involved in the work on the basis of outsourcing, work contracts and other tools.

Let us consider in more detail the model for assessing the impact of key technological competencies on the effectiveness of advanced technologies on the example of helicopter construction. To identify parts, units and machines, the development of which requires new competencies and building innovative technologies on their basis, it is advisable to use the method of functional and cost analysis, which will minimize the cost of purchasing the corresponding components of subcontractors. Functional cost analysis, discussed in detail in the work[11], is based on the use of a functional approach, which consists in considering the product as a set of functions that it should perform. Each of them is analyzed from the point of view of principles and possible ways of execution through the integration of special techniques. Assessment of options of the formed characteristics of helicopter components is carried out according to the criterion takes into account the degree of performance and importance of the functions and costs associated with ensuring that the helicopter equipment reaches a given level of performance.

The strength points of this method are the following:

---

11 Senderov V. L., Yurchenko T. I., Vorontsova Yu. V. Research of cost management tools and recommendations for their effective use in the management of the organization's economy // Energosberezheni y vodopodgotovka.– 2013. – № 2 (82). – Pp. 55-67.

- consideration of helicopter technology (a component of helicopter technology) as a set of functions that it should perform;
- coverage in the research process of all stages of the helicopter creation – from the moment of its design to the moment of its operation.

As a special form of functional cost analysis, the Target Costing – TC method can be considered, which is focused on ensuring the achievement of the specified cost indicators of a product at the design stage. When using this method, it is recommended to perform calculations in three stages:

- *identifying actions and their grouping:*
1. by the scale of production;
2. by ensuring the functioning of an organization as a whole. It is also recommended to classify resources for optimal cost analysis to form a competitive helicopter price. The division of resources makes it possible to organize a simple system of periodic cost reports solving both financial and managerial tasks;
- *identifying cost drivers* – the costs most sensitive to impact for effective management of key competencies, and defining indicators that characterize each of the resources;
- *calculating the cost of a helicopter,* taking into account the elimination of unnecessary and duplicate functions.

The method is very promising in terms of creating an information and regulatory framework. In addition, any changes in the internal environment of the organization will require a revision of this base, for example, cost drivers (their composition and relationship with resources). All of the above justifies the use of functional-cost analysis to solve not only cost management problems, but also to manage any types of resources both within an organization as a whole, and by individual divisions (production, functional), types of helicopter technology, including various stages of its creation.

The scheme of functional and cost analysis in an organization when designing components of helicopter equipment is shown in Figure 2.4.

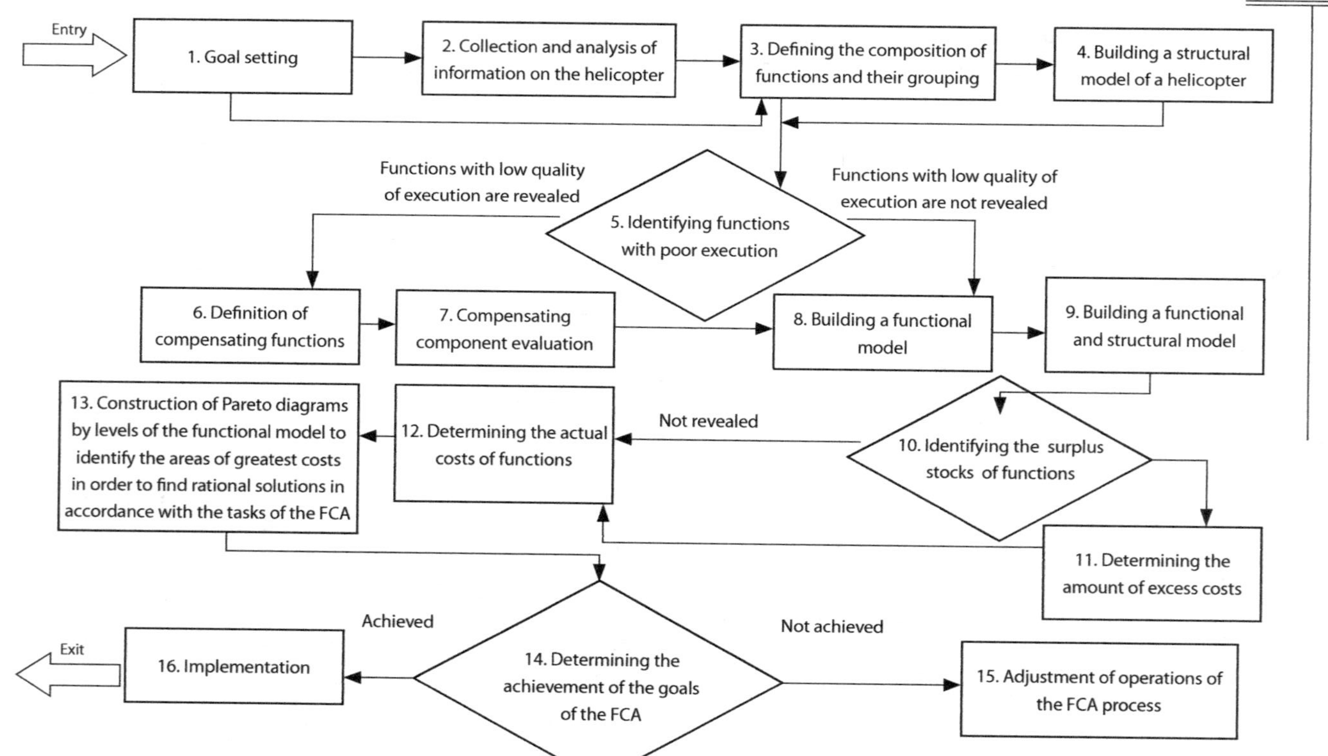

Figure 2.4. Scheme of functional cost analysis (FCA)

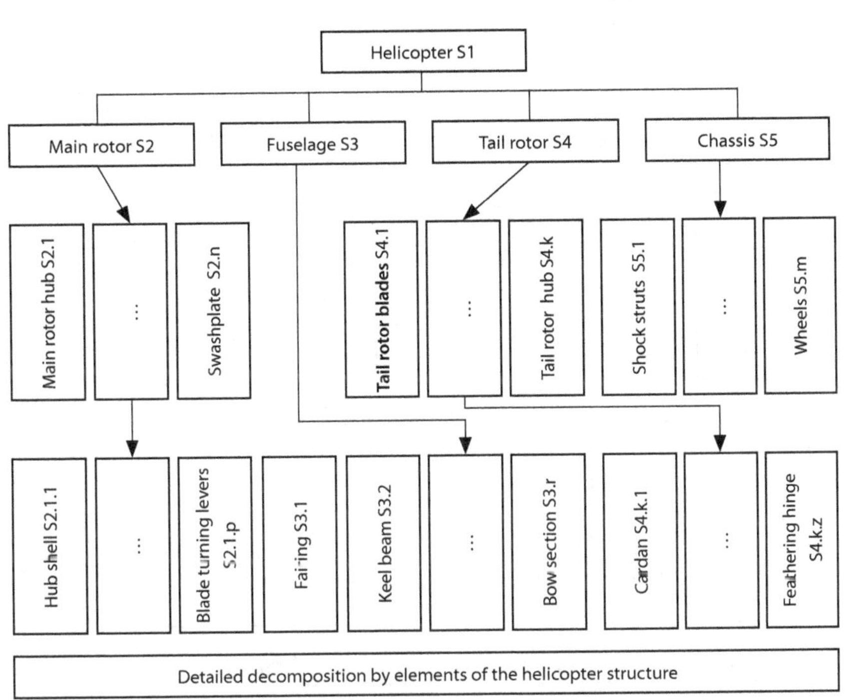

Figure: 2.5. Fragment of the structural model of a helicopter
for the functional and cost analysis

The following are steps for conducting a functional cost analysis.
Step 1. Building a structural model of a helicopter, shown in Fig. 2.5.
Step 2. Building a functional model for creating a helicopter.

At this stage, it is necessary to form a list of all the functions carried out by an organization in the process of designing a helicopter. A fragment of such a model is shown in Fig. 2.6.

Step 3. Building a functional and structural model.

In a functional cost analysis, in order to form a certain given level of competitiveness, it is necessary to build a functional structural model with reference to the cost of each function connected to the creation of each element of the helicopter. As most of a helicopter's elements are assembled from purchased components, certain functions are needed to assemble them into a coherent whole. A hypothetical functional and structural model is presented in table 2.1.

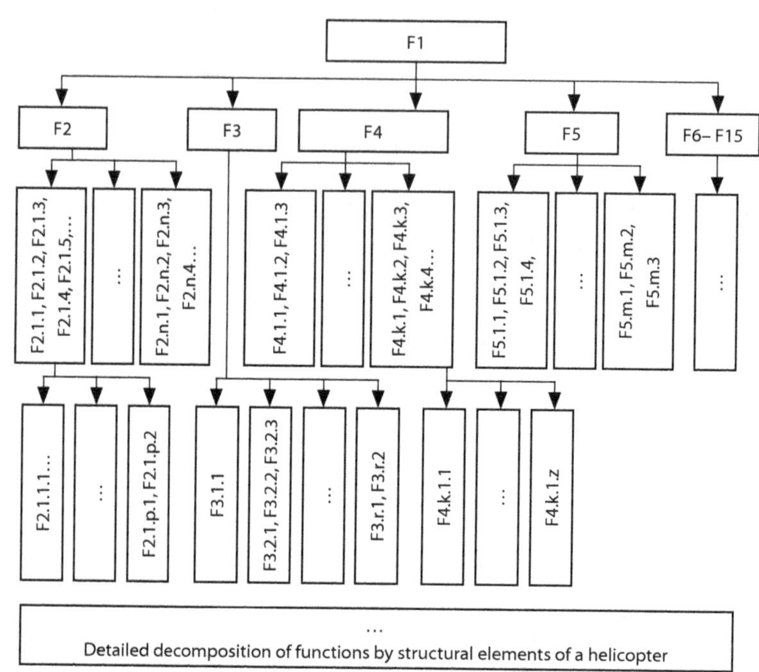

Figure 2.6. Fragment of a functional model of a helicopter for performing functional cost analysis

Table 2.1. Illustration of the construction of a functional-structural model for functional-cost analysis

	S1	S2	S3	S4	S5	S2.1	⋮	S2.n	S3.1	S3.2	⋮	S3.r	S4.1	⋮	S4.k	S5.1	⋮	S5.m	S2.1.1	⋮	S2.1.p	S4.k.1	⋮	S4.k.z	total costs
F1	+				v				v	v								v	v						
F2		+																							
F3		+																							
F4		+			v	v		v	v	v															
F5			+																						
F6			+																						
⋮																		v	v						

	S1	S2	S3	S4	S5	S2.1	⋮	S2.n	S3.1	S3.2	⋮	S3.r	S4.1	⋮	S4.k	S5.1	⋮	S5.m	S2.1.1	⋮	S2.1.p	S4.k.1	⋮	S4.k.z	total costs
F10				+				v	v																
F11					+																				
F12					+										v	v									
F13					+																				
F14	v	v	v		+																				
F15					+																				
F2.1.1						+																			
F2.1.2						+						v	v												
F2.1.3						+												v	v						
F2.1.4						+																			
F2.1.5						+						v	v								v	v			
⋮																									
F2.n.1		v	v			+																			
F2.1.1.1	v	v	v	v	v	v													+						
⋮																									
F2.1.p.2																					+				
F4.k.1.1																						+			

In Table 2.1, the cost of performing a specific function when creating and/or assembling each of a helicopter's structural elements is indicated by a "plus". However, due to the complexity and laboriousness of building a functional-structural model, the point that the same functions implemented in an organization can be performed in relation to various structural elements of the helicopter may be ignored. In this case, costs also arise, which are displayed in the functional-structural model with a "tick".

Step 4. Ranking the total cost of functions in descending order.

Step 5. Finding the specific weight of expenditure for each function in the total cost:

$$\text{Specific weight } E_F = \frac{E_F}{\sum_{F=1}^{z} E_F},$$

where E_F is total costs for the implementation of a specific function for all structural elements of the helicopter design.

Step 6. Calculation of costs for all functions on a cumulative basis.

Step 7. Grouping functions into A, B and C.

This grouping is performed as follows: those functions that make up 80% of the cumulative total when costs increase can be attributed to group A; those that make up from 80-90% of the cumulative total – to group B; those that make up more than 90% of the cumulative total – to group C. As a rule, activities for the development of key competencies are necessary when implementing the functions that fall into group A.

This analysis will allow finding unnecessary and/or duplicate functions, the rejection of which will give significant cost savings, and build relationships with related parties on favorable terms for the parent organization.

The proposed approach demonstrates that competencies allow increasing the rate of creation of new competitive advantages of products, which in modern conditions is a vital condition for the competitiveness of helicopter products and their components. Using mathematical modeling, it is possible to determine the optimal strategies for expenses for the development of new competencies, which will determine the creation of high techno-economic characteristics of products.

The process of creating competitive products requires increased competence and a careful approach to all components and aggregates of the created products. That is why it is important for high-tech industries to predict the needs of organizations and society as a whole and develop relevant key competencies and innovative technologies that determine the competitive advantages of future products.

2.3. METHODS FOR ASSESSING THE ACHIEVABILITY OF TECHNO-ECONOMIC PRODUCT INDICATORS THROUGH THE USE OF INNOVATIVE TECHNOLOGIES

At the early stages of the design of highly competitive products, analysis of innovative technologies is of particular importance, due to which products with promising techno-economic image will be developed. Such analysis can be carried out on the basis of methods for determining the basic indicators of competitiveness, set by technical characteristics that allow predicting the labor, material and cost of manufacturing a product at the early stages of design.

To develop a product with competitiveness indicators that meet the requirements of the technical specification, it is necessary to monitor these indicators at all stages of design development (development of an advanced pilot project, preliminary design, prototype, etc.).

Technical indicators of products are the source of their competitiveness. They require specification depending on the specifics of its purpose and the conditions of consumption and operation.

Let us consider the impact of innovative technologies on the competitiveness of high-tech products. Assessment of such an impact is an urgent task in the pre-project period, which makes it possible to form an effective scheme for managing the design process.

Innovative technology refers to a set of innovative solutions that improve the techno-economic characteristics of products.

Let us present the following innovative technologies, the introduction of which can increase the competitiveness of products:

$$I_1, I_2, \ldots, I_N.$$

Each considered innovative technology should work to improve the characteristics of products in the event of a successful introduction. Herewith, specific innovative solutions can be part of various innovative technologies. As a rule, innovative technology can improve several techno-economic characteristics of products. Therefore, to designate this property of innovative technology, we will use the vector of product characteristics:

$$Q(I_i) = \begin{pmatrix} q_1 \\ q_2 \\ \vdots \\ q_K \end{pmatrix}.$$

We assume that the innovative technology improves the various characteristics of the product.

Each innovative technology is described by the cost of the development and introduction the innovative technology. We take this value as

$$V = V(I_i).$$

In various applications of the economic and mathematical model under consideration, we will use the interval value as the V value, since in the conditions of high-tech enterprises the cost of the development and introduction innovative technologies may differ from the planned cost.

The use of any new technology, and especially innovative one, is fraught with risk. The risk in the introduction of innovative technologies consists in the failure to achieve the expected effect from their application and is measured by the value

$$0 \leq R(I_i) \leq 1.$$

At the same time, a zero value of the $R(I_i)$ value means a zero effect from the use of innovative technology, and a single value shows that the expectation of efficiency from the introduction of innovative technology has been fully justified.

Thus, the innovative technology will be described as follows:

$$I_i = \begin{pmatrix} Q_i \\ V_i \\ R_i \end{pmatrix}.$$

This vector will be used in the economic and mathematical model for choosing the optimal innovative technology that ensures the competitiveness of the products.

Another object of the model is the competitiveness of products. As already noted, this concept is a complex category, which should include a large number of factors. In the economic and mathematical model, we will consider the competitiveness of products as a vector consisting of numerical indicators of the competitiveness of these products. The set of these indicators is determined by the initial data for calculations.

Let us denote the considered competitiveness

$$CQ(I) = F(Q) = F \begin{pmatrix} q_1 \\ q_2 \\ \vdots \\ q_K \end{pmatrix},$$

where F is some function that connects competitiveness and the techno-economic characteristics of products improved through innovative technologies.

Thus, we consider K different indicators of competitiveness. In this case, we will assume that the indicator $CQ_i(I')$ corresponding to innovative technology is better than this indicator for innovative technology if the following inequality is satisfied

$$CQ(I') > CQ(I'').$$

Of course, only indicators of the same type can be compared. In this case, only a partial order can be introduced on the set of vectors of competitiveness indicators, since not all vectors can be compared.

Next, we formulate an economic and mathematical model for choosing the optimal innovative technology that ensures the competitiveness of products, as follows:

$$MQ = \max_{I_{i_1}, I_{i_2}, \dots, I_{i_L}} = (CQ(Q_{j_1}, Q_{j_2}, \dots, Q_{j_S}), R).$$

In this model, MQ is the indicator of the effective competitiveness of products, which shows the maximum efficiency of innovative technologies to increase the competitiveness of products, taking into account the cost of its development and introduction and the resulting risks.

This model does not reveal the functional relationship between some indicators of the competitiveness of products and individual innovative technologies. These functional dependencies and their mathematical description are specified when constructing specific economic and mathematical models.

Let us consider a model for assessing the techno-economic indicators of products and their competitiveness, formed under the influence of innovative technologies. Its algorithm is shown in Fig. 2.7.

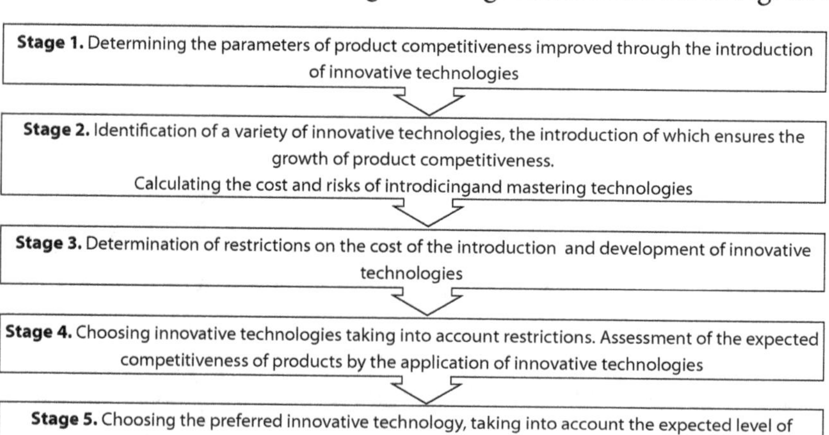

Figure: 2.7. Model for choosing the optimal innovative technology that ensures the competitiveness of the products

The parameters of the competitiveness of products, which are improved through the introduction of innovative technologies, are determined at the stage of forming the techno-economic image of the product. The techno-economic of products we refer to is a set of techno-economic indicators that determine the consumer properties of products.

At the stage of formalizing a variety of innovative technologies, it is necessary to assess the cost of introducing and mastering innovative technologies, as well as the risks of their introduction. Cost estimation can be performed by traditional methods, such as cost-based or comparative. Risk assessment of the introduction of innovative technologies is based on the methods of simulation of economic and mathematical modeling through discrete study of external and internal factors and events.

Determination of restrictions on the cost of introduction and development of innovative technologies is based on an analysis of the financial capabilities of an enterprise. Herewith, restrictions on the cost of introducing and mastering innovative technologies can be regulated by various documents that regulate the innovative development of an organization.

At the next stage, the selection of innovative technologies is carried out taking into account the constraints. From the entire set of technologies under consideration, at this stage it is necessary to choose those that will lead to an increase in the competitiveness of products while respecting cost constraints on their development and implementation.

The selected innovative technologies, the development of which falls within the previously defined financial constraints, are divided into several groups (clusters) in order to make the optimal choice of innovative technology. Let us divide a lot of innovative technologies, for example, into two clusters. This will allow grouping the most effective innovative technologies in one of the clusters.

Let us describe the procedure for selecting innovative technologies based on the clustering method.

Let us consider the following clustering problem: break down an existing set of N innovative technologies

$$I = \begin{pmatrix} I_1 \\ I_2 \\ \vdots \\ I_N \end{pmatrix}$$

into two clusters based on the principle of similar impact on the competitiveness of products.

Under the cluster we will understand a set of objects, on the multitude of which the following characteristics common to the whole set are analyzed:

- the cluster center. Mathematically, a cluster center is the average locus of points in the considered space of variables (in our case, in the space of innovative technologies);
- ability to determine the cluster radius. It represents the maximum possible distance of points from the center of the cluster. In our case, the cluster radius is the maximum deviation of the innovative technology characteristics from the average values for this cluster;
- calculated cluster size. This is the number of points located within the cluster. In the context of our problem, the cluster size will be the number of innovative technologies whose characteristics are separated from the cluster center by a distance less than the cluster radius.

The fundamental component of the procedure for solving cluster analysis problems is the choice of the scale of the parameters that make up the optimality criterion. Often, some parameters can take on rather small numerical values, while others, on the contrary, are large. In this case, the influence of parameters that take large values will be significantly higher than parameters that take small values. Another problem is that the growth of some parameters can lead to an increase in the competitiveness of products, while the growth of others leads its decrease. For such variables, cluster analysis provides for a preliminary procedure for standardizing the variables, for example, using the normalization method.

A necessary element of cluster analysis is a quantitative assessment of the proximity of two objects, i.e. calculating the distance between them. Let the proximity of two objects Q, Q^* (vectors of competitiveness indicators) be given by d_{QQ^*}. This measure satisfies the following properties:

$$d_{QQ^*} > 0, \ d_{QQ} = d_{Q^*Q^*} = 0,$$

$$d_{QQ^*} = d_{Q^*Q},$$

$$d_{QQ_1} + d_{Q_1Q^*} \geq d_{QQ^*},$$

where Q_1 is some arbitrary object from the set.

As the main metric for determining the distance between two vectors of competitiveness indicators, we will use the formula corresponding to the Euclidean distance:

$$d_{E_{ij}} = \sqrt{\sum_{l=1}^{m}\left(x_i^l - x_j^l\right)^2} \qquad (2.1)$$

The criterion of optimality in clustering objects is a sufficient value of the measure of proximity of the vector of competitiveness indicators obtained as a result of the use of innovative technology in the production and technological cycle to the vector of reference values of competitiveness indicators.

To divide innovative technologies into groups according to the degree of their impact on the competitiveness of products, we use the method of hierarchical clustering. This method involves sequentially combining smaller clusters into larger ones or dividing larger clusters into smaller ones. We will use the agglomerative clustering method. It assumes that each object is a separate cluster, indicated in Fig. 2.8 as A, B, C, and D, and then aggregates into larger clusters (Figure 2.8).

The algorithm for dividing the set of innovative technologies into two clusters involves a two-stage procedure.

1. Selection of two a priori objects as initial cluster centers.

At this stage, it is necessary to estimate the distances between all the competitiveness vectors corresponding to innovative technologies in pairs using the formula (2.1). As two a priori objects,

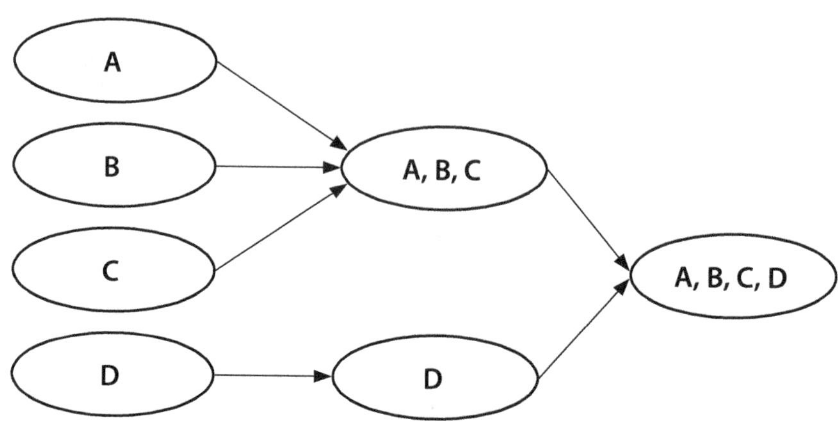

Figure: 2.8. Agglomerative clustering method

a pair of innovative technologies should be selected with the greatest distance between them.

2. Consistent attribution of innovative technologies to one of the clusters.

At this stage, innovative technologies are sequentially sorted out. For the selected innovation technology, the distance from the corresponding competitiveness vector to the cluster center is determined. Innovative technology should be attributed to the cluster, with the distance to the center. After a sequential search of all innovative technologies, two groups of innovative technologies will be obtained.

For further selection of the optimal innovative technology, it is necessary to define the concept of the integral indicator of the competitiveness of products, described by the vector of competitiveness indicators[12].

The coefficient of competitiveness K_i of products according to the selected techno-economic characteristics can be calculated by the formula

$$K_i = \frac{q_i}{q_i^0} \, ,$$

---

12 Chursin A.A. Theoretical foundations of competitiveness management: theory and practice. – M .: Spektr, 2012 .-520 p.

where is the value of the i-th private indicator of the competitiveness of the analyzed products from the set of private indicators (components of the vector Q); is the value of the i-th private indicator of the competitiveness of competing products from the set of its private indicators.

In the case of considering a set of competing types of products, the coefficients of competitiveness of products for a particular characteristic should be determined in comparison with the leading value of the characteristic from the set of characteristics of analogous products.

Next, it is necessary to compare private indicators of product competitiveness with indicators that reflect the achievements of scientific and technological progress for this type of product. The competitiveness coefficient is calculated by the formula

$$K'_i = \frac{q_i}{q'_i} \, ,$$

where is the value of the i-th private indicator of competitiveness from the set of private indicators reflecting the achievements of scientific and technological progress.

The generalized indicator of product competitiveness is calculated as follows:

$$CQ = \sum_{i=1}^{m} \alpha_i (K_i + K'_i) ,$$

where m is the number of counted private indicators; are weights reflecting the importance of each competitiveness indicator

$$\sum_{i=1}^{m} \alpha_1 = 1 .$$

The proposed generalized indicator of the competitiveness of products allows comparing different types of products by the level of competitiveness.

We will select one of two clusters, where innovative technologies will have a more significant impact on building product competitiveness. To do this, we calculate generalized indicators of competitiveness with respect to the vectors Q of competitiveness corresponding to the cluster centers. Further selection of the optimal innovative technology will be carried out within the cluster, the center of which corresponds to the best value of the generalized indicator of competitiveness.

The final selection of innovative technology, the use of which will ensure a sufficient level of product competitiveness, will be made taking into account the risk level.

The selection of innovative technologies at this stage is carried out from a variety $I_1, I_2, ..., I_S$ of innovative technologies that make up the selected cluster. For each innovative technology, we define a generalized indicator of competitiveness CQ_i, where $i = 1, 2, ..., S$. Also, for innovative technologies, the risks R_i, were identified, where $i = 1, 2, ..., S$, related with the introduction of innovative technologies.

The final selection of innovative technology will be based on the calculation of the effective competitiveness indicator MQ, which shows the maximum efficiency of innovative technologies to increase the competitiveness of products Based on the previous model, the MQ indicator can be calculated using the following formula:

$$MQ = \max_{I_1, I_2, ..., I_S} (CQ_i \cdot (1 - R_i)),$$

where $i = 1, ..., S$.

Thus, the optimal innovative technology is the one that will ensure the greatest increase in the competitiveness of products, taking into account dynamically changing factors and risks. At the same time, the cost of developing and mastering such technology will not go beyond the previously defined border.

The proposed model for assessing the techno-economic indicators of products and their competitiveness, formed under the influence of innovative technologies, allows us to determine the techno-economic indicators of products, which, can be achieved as a result of the development and introduction of innovative technologies and will ensure sales products on the market.

Let us consider an example for the application of the proposed model for assessing the techno-economic indicators of products formed under the influence of innovative technologies. To do this, we will analyze the creation of the Mi-38 multipurpose medium-range helicopter, developed by the Russian Helicopters holding. When creating this helicopter, a number of innovative technologies were applied in order to form breakthrough products and significantly increase the competitiveness of this product at the domestic and global levels. We will consider some of them in table. 2.2.

The Mi-38 helicopter was developed within the Federal Target Program "Development of Civil Aviation in Russia for 2002–2010 and for the Period until 2015"[13]. It should be noted that modeling the impact of innovative technologies on competitiveness can also be carried out at the pre-project stage.

Table 2.2. Innovative technologies

№	Brief description of the technolog	Improved features
1	Application of elastomeric bearings in the main rotor hub	Extending the bushing life
2	Use of additional damping chambers in the helicopter landing gear and damping devices in the pilot and passenger seats	Improving security settings
3	Changing the position of the power plant. Assignment of the PU for the helicopter axis, for the main gearbox	Increased comfort – reduced noise and vibration
4	The use of fiberglass materials in the manufacture of rotor blades and tail rotor blades	A significant increase in the resource. Reducing the specific labor intensity of maintenance
5	Application of the latest electronic equipment	Crew reduction at the expense of the flight engineer
6	Using the technology of manufacturing a screw by winding a carbon fiber tape on a rotating tooling	Reduced product cost and improved screw stability

The use of these innovative technologies, as well as many others, allowed the holding to create competitive products.

Table 2.3 presents the performance characteristics of the Mi-38 helicopter.

---

13 A.A. Kotlyarov Development of the Mi-38 Innovative Competitive Helicopter// Innovations. – 2013. –No. 12. – Pp. 71–76.

Table 2.3. Flight performance of the Mi-38 helicopter

№	Parameter	Indicator
1	Main rotor diameter, m	21,10
2	Tail rotor diameter, m	3,84
3	Length, m	25,22
4	Height, m	5,56
5	Width, m	4,50
6	Empty weight, kg	8300
7	Normal takeoff weight, kg	14200
8	Maximum takeoff weight, kg	15600
9	Engine's type	2GTE-PW-127 T/S
10	Power, h.p.	2x2500
11	Maximum speed, km / h	320
12	Cruising speed, km / h	290
13	Ferry range, km	1350
14	Practical range, km	820
15	Radius of action, km	325
16	Practical ceiling, m	5200
17	Static ceiling, m	2500
18	Crew, people	2
19	Payload	32 passengers or 5000 kg of cargo in the cabin or up to 6000 kg of cargo on the suspension

Source[14].

Since the exact description of the features of innovative technologies used in the development of the Mi-38 helicopter is sensitive information, we will use model information in the calculations.

Phase 1. Determination of the parameters of the competitiveness of products, improved through the introduction of innovative technologies.

To assess the competitiveness of the Mi-38 helicopter, it is necessary to compare its features with existing analogues developed by foreign helicopter companies. For this comparison, we will use table. 2.4, which shows an example of a set of competitiveness indicators for the Mi-38 (Russian Helicopters), H-225 (Airbus

---

14 A.A. Kotlyarov Development of the Mi-38 Innovative Competitive Helicopter// Innovations. – 2013. –No. 12. – Pp. 71–76.

Helicopters), S-92 (SikorskyAircraft), AW-101UT (AgustaWestland) models.

Table 2.4. Techno-economic specifications of helicopter equipment that form its competitiveness

№	Specification	Mi-38	H-225 Airbus Helicopters	S-92A Sikorsky Aircraft	AW-101UT Agusta Westland
1.	Maximum takeoff weight, kg	16200	11200	12837	15600
2.	Payload weight, kg	6000	5624	4955	6000
3.	Static ceiling, m	2450	795	2042	1460
4.	Cruising speed, km / h	290	262	280	278
5.	Flight range with full refueling of fuel tanks, km	1200	838	1013	1440
6.	Helicopter price, USD million	17,6	27,9	26,9	29,0

Source: compiled by the author based on.[15]

According to the table. 2.4 most of the specifications of the Mi-38 helicopter are completely superior to those of competing helicopters. Based on this example, let us evaluate the impact of innovative technologies on the competitiveness of products.

Phase 2. Identification of a variety of innovative technologies, the introduction of which ensures the growth of product competitiveness. Calculation of the cost and risks of introduction and mastering technologies.

Let us assume that in he given example, when creating the Mi-38 helicopter, they used innovative technologies, which we will denote as I_1, I_2, I_3, I_4. According to our approach, each innovative technology is described, on the one hand, by the expected effect with an increase in a given flight performance, on the other hand, by a certain risk when using this technology. For the sake of certainty, we will assume that the considered innovative technologies allow us to improve the flight characteristics presented in table 2.5.

15 A.A. Kotlyarov Development of the Mi-38 Innovative Competitive Helicopter// Innovations. – 2013. –No. 12. – Pp. 71–76.

Table 2.5 Flight performance

№	Improved specifications
1	Maximum takeoff weight
2	Payload mass
3	Static ceiling
4	Cruising speed
5	Flight range with full fuel tanks

In dimensionless quantities, we will set the expected cost of development and introduction of an innovative technology and the possible risk. To do this, consider the corresponding vectors

$$I_j = \begin{pmatrix} V_j \\ R_j \end{pmatrix},$$

where V_j is the expected cost of development and introduction of innovative technology; R_j is the level of risk associated with the implementation of this technology.

In our example, we will use the following model values

$$I_1 = \begin{pmatrix} 2.5 \\ 0.2 \end{pmatrix},$$

$$I_2 = \begin{pmatrix} 1.9 \\ 0.3 \end{pmatrix},$$

$$I_3 = \begin{pmatrix} 2.1 \\ 0.25 \end{pmatrix},$$

$$I_4 = \begin{pmatrix} 1.7 \\ 0.15 \end{pmatrix}.$$

Phase 3. Determination of restrictions on the cost of introducing and mastering innovative technologies I_2, I_3, I_4.

Let the cost of development and introduction of innovative technology not exceed 2.2 conv. units established by governing documents. Thus, the innovative I_1 technology does not satisfy the limitation on the marginal cost of its development and implementation. Thus, further selection of innovative technologies is made among technologies I_2, I_3, I_4.

Phase 4. Selection of innovative technologies given constraints. Assessment of the expected competitiveness of products as a result of the application of innovative technologies.

The selection of innovative technologies that best affect the competitiveness of helicopter technology will be performed using the agglomerative clustering method. To do this, we will select two a priori objects around which clusters will form. For this, it is necessary to calculate the distances between the vectors of the parameters of the competitiveness of products achieved as a result of the application of innovative technologies. When calculating the distances, we will be guided by the model table of the initial data (Table 2.6).

Table 2.6. Model table of initial data

Indicator	Innovative technologies		
	I_2	I_3	I_4
Maximum takeoff weight, kg	16200	15800	15000
Payload weight, kg	6000	5000	5500
Static ceiling, m	2450	2500	2200
Cruising speed, km / h	290	200	220
Flight range with full refueling of fuel tanks, km	1200	1150	1100

The parameters of the norm of indicators are presented in table 2.7 (the parameters are normalized so that the sum is one).

Table 2.7 Indicator norm parameters

Indicator	Innovative technologies		
	I_2	I_3	I_4
Maximum takeoff weight, kg	0,345	0,336	0,319
Payload weight, kg	0,364	0,303	0,333
Static ceiling, m	0,343	0,350	0,308
Cruising speed, km / h	0,408	0,282	0,310
Flight range with full refueling of fuel tanks, km	0,348	0,333	0,319

Source: compiled by the author based on[16].

The distances calculated for the normalized vectors are as follows:

$$d_{I_2 I_3} = 0,1,$$

$$d_{I_2 I_4} = 0,21,$$

$$d_{I_3 I_4} = 0,13.$$

The greatest distance is observed between the vectors corresponding to the second and fourth innovation technology. Thus, clusters will be formed around these innovative technologies.

Since an innovative I_3 technology must be assigned to one of the clusters, it is necessary to estimate the distance from the corresponding vector of competitiveness indicators to the cluster centers:

$$d_{Cluster\ 1,\ I_3} = 0,1,$$

$$d_{Cluster\ 2,\ I_3} = 0,13.$$

Due to the fact that the distance to the first cluster is less, then this innovative technology should fall into the first cluster.

To select a cluster whose innovative technologies are more effective in terms of influencing the competitiveness of the helicopter, we will assess the generalized competitiveness indicators based on the vectors of competitiveness indicators corresponding to the cluster centers.

Since the second cluster consists of one innovative technology, the corresponding vector will be the center of the cluster.

For the first cluster, the coordinates of its center can be defined as the geometric mean of the corresponding coordinates of the vectors of competitiveness of innovative technologies located in the cluster. In this case, it is necessary to obtain a vector of unnormalized indicators

---

16 A.A. Kotlyarov Creation of an innovative competitive product – the Mi-38 helicopter // Innovations. – 2013. – No. 12. – Pp. 71–76.

$$Q_{Cluster1} = \begin{pmatrix} 15998 \\ 5477 \\ 2474 \\ 240 \\ 1174 \end{pmatrix},$$

The initial data for calculating the indicators of the competitiveness of cluster centers are shown in table. 2.8.

Table 2.8. Initial data for calculating the indicators of the competitiveness of cluster centers

Parameter	Cluster center 1	Cluster center 2	The best indicator among the market leaders H-225, S-92A, AW-101UT	Level of scientific and technical development for products (helicopters weighing 10-16 tons)
Maximum takeoff weight, kg	15998	15000	15600	18000
Payload weight, kg	5477	5500	6000	7000
Static ceiling, m	2474	2200	2042	3500
Cruising speed, km/ h	240	220	280	320
Flight range with full refueling of fuel tanks, km	1174	1100	1440	1500

Source: compiled by the author based on [17].

Let us assume that the indicators of competitiveness are in equilibrium. In this case, the generalized indicators of competitiveness corresponding to the cluster centers are calculated as follows:

---

17 A.A. Kotlyarov Creation of an innovative competitive product – the Mi-38 helicopter // Innovations. – 2013. – No. 12. – Pp. 71–76.

$$CQ_{Cluster\,1} = \left(\begin{array}{l} 0,2 \cdot \left(\dfrac{15998}{15600} \right) + 0,2 \cdot \left(\dfrac{5477}{6000} \right) + \\[2mm] +0,2 \cdot \left(\dfrac{2474}{2042} \right) + 0,2 \cdot \left(\dfrac{240}{280} \right) + 0,2 \cdot \left(\dfrac{1174}{1440} \right) \end{array} \right) = 0,964.$$

$$CQ_{Cluster\,2} = 0,2 \cdot \left(\frac{15000}{15600} \right) + 0,2 \cdot \left(\frac{5500}{6000} \right) +$$

$$+0,2 \cdot \left(\frac{2200}{2042} \right) + 0,2 \cdot \left(\frac{220}{280} \right) + 0,2 \cdot \left(\frac{1100}{1440} \right) = 0,901.$$

Based on the comprehensive assessments of competitiveness, we conclude that the final selection of innovative technologies should be carried out among the innovative technologies that form the first cluster.

Phase 5. Selection of the preferred innovative technology, taking into account the expected level of competitiveness and risk.

We will calculate a comprehensive assessment of the competitiveness of helicopters, the techno-economic parameters of which are achieved as a result of the application of innovative technologies I_2 and I_3. The calculation is carried out similarly to the above procedure for clusters. As a result, we obtain the following values of the generalized indicator of competitiveness:

$$CQ(I_2) = 1,23,$$
$$CQ(I_3) = 1,02.$$

The calculation of the effective competitiveness indicator MQ, which shows the maximum effectiveness of innovative technologies to increase the competitiveness of products, will be carried out on the basis of the calculated values of generalized competitiveness indicators and risk level values:

$$MQ = \max_{I_2,I_3}\left\{ (1,23 \cdot (1-0,3)),\, (1,02 \cdot (1-0,25)) \right\} =$$

$$= \max_{I_2,I_3}\left\{ 0,861;\, 0,816 \right\} = 0,861.$$

The resulting value corresponds to the use of innovative technology I_2, which achieves the techno-economic indicators of the Mi-38 helicopter, presented in table. 2.4 The economic meaning of the MQ value is a quantitative indicator of the Mi-38 helicopter in competition with similar models produced by foreign companies. It can be concluded that when choosing the optimal innovative technology that ensures the competitiveness of the products of the Russian Helicopters holding – the Mi-38 helicopter, the considered innovative technologies have enabled to achieve the techno-economic parameters, which allowed domestic helicopter industry to produce competitive products, provide sales helicopter on the market.

Thus, the assessment of the impact of innovative technologies on the competitiveness of products is a critical step of the pre design period, when the techno-economic image of a product is determined and the amount of resources required is estimated. The choice of innovative technologies is made on the basis of the obtained estimates due to which the competitive advantages of the final product will be achieved taking into account market and consumer expectations.

The mathematical model is demonstrated on a model example related to the assessment of the impact of innovative technologies on the competitiveness of the products of the Russian Helicopters holding – the Mi-38 helicopter. The given example shows the efficiency of the proposed model, which can become a theoretical basis for developing a methodology for assessing the effectiveness of the introduction of innovative technologies. Also, the results obtained can be used to develop information and analytical systems to manage the competitiveness of products.

CHAPTER 3. ECONOMIC TOOLS FOR PLANNING RESOURCE LIFE CYCLE SUPPORT FOR PROMISING HIGH-TECHNOLOGY PRODUCTS

3.1 METHODS FOR PLANNING THE COST OF RESEARCH AND DEVELOPMENT WORK

When developing, creating or upgrading complex technical systems, the most important task is to assess the resource support required for R&D, as well as the relevance of the calculated estimates. Particular attention is paid to the assessment of the accuracy of determining financial resources for R&D, since work in the design of complex technical systems involves significant financial costs. For example, in helicopter engineering, the main problem is that the development of helicopter equipment has fundamental features, since many units, machines and systems of helicopters are unique developments based on the wide use of innovative technologies. In this regard, in the course of R&D, it often turns out that there is a single developer of specific units, machines or systems (the developer can be both the parent organization and a subcontractor), therefore it is necessary to assess the justification of the financial resources required to successfully solve a specific problem within the R&D (including work performed by a subcontractor). Another feature of R&D for the development of helicopter equipment is the stage of flight tests, and forecasting their number is a separate task that requires special research.

Using the tools proposed in the work, it is possible to obtain a relevant assessment of the resources for R&D. Its use should ensure the effectiveness of R&D for the development of helicopter equipment, saving financial resources and all types of other resources in the complex.

Assessment of the amount of financial resources required to carry out R&D work on the development of helicopter equipment can be carried out by various methods, including standard methods for justifying the cost of R&D, but such a justification has its own characteristics due to the specifics of the works in the development of complex technical systems. The main ones include the absence of a competitive market for basic parts, assemblies and assemblies, and consumers are mainly organizations that are part of the «Russian Helicopters» JSC holding. Therefore, when calculating the resourcing, it is difficult to use methods related to the analysis of the cost of similar developments. In addition, the developed units, machines and systems are unique, which also complicates the use of statistics on the cost of R&D development.

Due to these circumstances, a methodology that allows justifying the resourcing for performing R&D on the development or modernization of helicopter equipment is an urgent task, which will give the opportunity to prove or disprove the correctness of the reasonable prices for the customer and make decisions about contracting.

At the current level, the development of parts, units and machines of helicopter equipment can include complex scientific projects and technically complex industrial products. In addition, the creation of special software plays an important role in the development and modernization of helicopter systems. At the same time, the high reliability of these means must be confirmed. The traditional reliability requirements for parts, units and systems are reflected in the technical specifications for their development, therefore, these factors should also be taken into account when justifying the R&D resourcing.

The purpose of the presented methodology is to conduct a multilevel examination that solves the problems related to relevance obtaining objective and subjective quantitative assessments of the justification of the resourcing. However, obtaining a quantitative assessment based on the available statistical material and using expert analysis is a critical issue.

To apply the methodology for justifying the resourcing required for R&D to develop helicopter equipment, the following initial data are used:

- a text of the justification for R&D resourcing;
- resources for various stages of R&D;
- total R&D costs;
- a statistical base with information on the performance characteristics of a prototype and resources for conducting R&D for its development.

As estimates of the volume of various types of resources, their value term is used.

The first stage is calculation of the average regression cost of conducting R&D based on statistical information about similar R&D conducted earlier, using the specified parameters of the prototype gained from R&D. As a result of this stage, an objective cost estimate is obtained, which at the second stage will be used to assess the relevance of a reasonable estimate of the resourcing volume.

Let the products created through of R&D be characterized by various parameters of performance characteristics x_1, x_2, x_{Np}.

The regression model is as follows. The volume of the required resource provision (in value terms) is used as an independent variable, and the parameters of performance characteristics are used as explanatory variables. The regression equations are as follows:

$$y_k = a_0 + a_1 x_k^1 + a_2 x_k^2 + \ldots + a_{N_P} x_k^{N_P} + \varepsilon_k,$$

$$k = 1, 2, \ldots, L,$$

where are the coefficients of the model; are normally distributed and pairwise independent regression residuals of a random nature.

In some cases, it makes sense to use a power regression model rather than a linear one:

$$y = a_0 x_1^{a_1} \cdot x_2^{a_2} \cdot \ldots \cdot x_{N_P}^{a_{N_P}}.$$

Such a regression model consists in the linear model discussed above using the logarithm:

$$\ln y = a_0 + a_1 \ln x_1 + a_2 \ln x_2 + \ldots + a_{N_P} \ln x_{N_P}.$$

The method proposed above for calculating the of R&D resourcing (in value terms) for the development of units, machines and systems of helicopter equipment is based on the use of multiple regression. However, pairwise regression is often used for calculations on the most important performance characteristics.

In the conditions of multicollinearity, the following algorithm for evaluating the resourcing for the development of units, machines and systems of helicopter equipment may be useful.

Step 1. Paired linear regression is used for each regressor. Let us suppose we get a cost estimate using the regressor, which we denote by y_i.

Step 2. For each regressor x_i, we calculate the corresponding coefficient of determination, which we denote by R_i.

Step 3. We calculate the weights for each estimate using the following formula:

$$\varrho_i = \frac{R_i}{R_1 + R_2 + \cdots + R_{N_P}}.$$

Step 4. We determine the integral assessment of R&D resourcing using the following formula:

$$y = \sum_{k=1}^{N_P} \varrho_k y_k.$$

Such an algorithm leads to the fact that the pair regression with a higher coefficient of determination makes a greater contribution to the estimation of resourcing.

The above mathematical model for assessing the functional dependence of R&D cost on performance parameters based on data on the cost of developing basic products (units, machines and systems) contains a large number of computational operations that can be automated using software systems.

Thus, at the first stage, we obtain the average regression estimate of the R&D cost. This estimate is used to calculate the relevance index for justifying the R&D resourcing according to the following scheme.

Let us calculate the relevance index for justifying the R&D resourcing, which will assess the relevance of the cost of R&D itself. Let us define the relative coefficient by the formula

$$\Delta = \frac{|D - S|}{S},$$

where D is the obtained R&D cost estimate calculated according to this method; S is the declared R&D cost for the development or modernization of helicopter equipment (units, machines, systems).

According to the values of this coefficient, the relevance index for justifying the R&D resourcing is calculated according to the following scheme:

if $\Delta < 0{,}1$, then $I_D = 5$;

if $0{,}1 \leq \Delta \leq 0{,}2$, then $I_D = 4$;

if $0{,}2 \leq \Delta < 0{,}3$, then $I_D = 3$;

if $0{,}3 \leq \Delta < 0{,}4$, then $I_D = 2$;

if $0{,}4 \leq \Delta < 0{,}5$, then $I_D = 1$;

if $0{,}5 \leq \Delta$, then $I_D = 0$.

At the second stage of the methodology, it is necessary to justify the costs of the parent organization to support the work performed by co-executors of R&D (related organizations).

When the parent organization conducts R&D on the development of helicopter equipment for performing various tasks in R&D with the involvement of third-party performers, certain costs arise from various services of the parent organization to support the work of co-executors. Since as a rule the actions of the head R&D contractor are planned in nature with a fixed staffing schedule, the actual task is to calculate and justify costs (labor costs, financial resources) to support the work of R&D co-executors.

The main problem in justifying costs is the occurrence of additional, previously unaccounted activities while supporting work by R&D co-executors. The reason for this is also the complexity and uniqueness of R&D carried out by co-performers, since the design of helicopter equipment is innovative both in Russia and in the world.

At the second stage of using the methodology for justifying the costs of supporting the work of R&D co-executors, a regression assessment of these costs is carried out depending on various economic parameters of the R&D itself, in particular, depending on the expenses when performing the R&D.

The proposed methodology makes it possible to justify the costs of the parent organization to support work by R&D co-executors. Such costs consist of various components, in particular, when ensuring the conduct of R&D by co-performers, as well as when accepting the results. Due to the complexity of R&D activities carried out in the interests of the parent organization, additional costs may arise on its part, which depend on the nature and economic parameters of these R&D activities.

To justify the additional costs that arise when supporting the work of R&D co-executors, this methodology offers various econometric methods related to the analysis of statistical information, as well as models for estimating costs based on objective economic parameters of R&D performed by third-party performers.

Algorithms for estimating the costs of supporting work by R & D co-executors.

The application of the methodology for justifying the costs of supporting the work of R&D co-performers, considered in the work, is carried out in three stages.

1. Identification of the main work with an estimate of the costs of supporting the work of R&D co-executors.
2. Regressive estimation of these costs.
3. Final justification of the costs of supporting the work of R&D co-executors.

Let us consider stage 1 – identifying the main work with an estimate of the costs of supporting the work of R&D co-executors.

At the first stage, it is necessary to estimate the costs of the parent organization to support the work of R&D co-executors based on a step-by-step list of costs that are needed to support the work of related R&D co-executors. Let us consider the steps of the algorithm for this stage.

Step 1. Definition of the main groups of works to support the work of R&D co-executors.

Typical costs here are the following cost groups:

- development of terms of reference (ToR) for an integral part of R&D;
- approval of ToR;
- calculation of the contract price of R&D constituent part (CP);
- preparation of tender documentation
- conducting competition;
- contracting with co-executors of R&D CP;
- carrying out accounting support for the work R&D CP;
- preparation of initial data for performers;
- consulting performers;
- analysis and expertise of R&D CP results;
- preparation of documentation for the completion of R&D CP;
- etc.

At this step of the algorithm, the following table of possible labor costs is compiled

Table 3.1. Table of labor costs

№	Job title	Number of repetitions	Execution cost
...	...	...	...
I	N_i	K_i	T_i
...	...	...	...

Step 2. Calculation of the total costs of supporting the work of R&D co-executors.

This calculation is carried out according to the formula

$$T = \sum_{i=1}^{N} K_i \cdot T_i.$$

In this formula, the value T is the total costs estimate for supporting of R&D co-executors.

Step 3. Calculation of the error for estimating total costs.

When calculating the costs required to support the work of R&D co-executors, the error estimate should also be used in calculating the total labor costs.

The error is estimated using the formula

$$\Delta T = \left(\frac{T}{aT + b} \right)^c.$$

It is proposed to use the following values as specific values of empirical coefficients: a = 2; b = 1; c = 2/3.

We shall not that when using this estimate for fixed coefficients, the error estimate increases as the value of T increases, which reflects the fact that for large values of labor costs, the error in estimating total costs increases.

Step 4. Justification of the costs of the parent organization to support the work of R&D performers.

Previously, the following values were obtained:

T – total costs;

Δ – estimation of cost estimation error.

An estimate of the labor costs of the head organization to support the work of R&D performers will be the value that meets the following estimate:

$$\left| T' - T \right| \leq T \cdot \Delta T.$$

At this stage of the methodology, the relevance of the justification of the R&D cost is assessed using the evaluation expertise method.In this case, a questionnaire is used, the answers to which help to calculate the relevance index for justifying the R&D resourcing. Since the questionnaire is filled in by an expert analyst, a natural inevitable subjective factor arises, as helicopter equipmentfalls into complex technical systems category, for which there may be insufficient objective data to conclude on the level of relevance for justifying the R&D resourcing

Since various methods are used in regulatory practice to justify the cost of R&D, when analyzing the relevance of the cost according to this methodology, it is necessary to separate the following groups of information:

- list of stages of R&D;
- list of the performance characteristics of the prototype obtained from R&D.

Let us denote by S the set of stages of R&D, and by X – the set of performance characteristics of a prototype.

Let us consider the algorithms for applying the methodology for assessing the justification of R&D resourcing. Methods based on market price comparison (market analysis), as well as the cost method, can be used here. Therefore, the methodology for assessing the relevance of the R&D cost applies two algorithms to assess the justification of the cost in both cases. *Algorithm for assessing the relevance for justifying the R&D resourcing based on the relationship of performance characteristics and costs.*

After assessing the the relevance for justifying the R&D resourcing for the development of helicopter equipment (its units, machines, systems, etc.), the next stage of the methodology is to examine the relationship between the performance characteristics of prototypes developed within R&D and costs. The main problem in analyzing the cost of R&D calculated by the cost method is the analysis of the correctness of costs. Let S denote the set of stages of R&D. For each stage

$$s_i \in S, \quad i = 1, 2, \dots M,$$

where M is the total number of stages in the implementation of R&D.

Next, the index of the feasibility of the stage is calculated using an expert survey based on the expert questionnaire matrix (Table 3.2). When answering, the experts use the appropriate scale [18]:

18 Shermadini M. V., Nazarenko A. M. Expert systems in decision-making in innovative activity of high-tech enterprises // Economics and management: problems, solutions. – 2018. – Vol. 2. – No. 8. – pp. 111-117.

0 – no answer from the justification text;
1 – completely irrelevant;
2 – rather relevant;
3 – possibly relevant;
4 – rather relevant;
5 – completely relevant.

Table 3.2. Expert matrix-questionnaire

Question	Answer of Expert 1	...	Answer of Expert N
Evaluate the degree of relevance of the impact of the stage on the performance characteristics of the prototype			
Evaluate the degree of relevance of the stage cost			

Based on expert assessments, the following values are calculated by averaging expert assessments: i-

$$X_i = \frac{1}{N}\sum_{j=1}^{N} x_{ij}, \quad i = 1, 2, ...M$$

to assess the degree of relevance of the impact of the i-th stage on the performance characteristics of the prototype in the implementation of R&D and

$$Y_i = \frac{1}{N}\sum_{j=1}^{N} y_{ij}, \quad i = 1, 2, ...M$$

to assess the degree of relevance of the cost of the i-th stage.

Further, these values are averaged according to the formula of the geometric mean

$$I_{S_i} = \sqrt{X_i \cdot Y_i}.$$

Using a geometric mean allows for a more accurate assessment of the impact of these estimates.

Finally, according to the methodology for assessing the relevance of justifying the cost of R&D for the development of helicopter equipment, the index of the relevance of the cost of R&D stages and their impact on the performance characteristics of the prototype is calculated. This index is determined by the arithmetic mean formula:

$$I_S = \frac{1}{M}\sum_{k=1}^{M} I_{S_k}.$$

After preliminary assessments of the relevance of the justification of the R&D cost, the general index of the relevance of the cost justification is calculated using the following algorithm.

Algorithm for index aggregation of relevance of R&D cost justification.

Based on preliminary calculations, we have the following values:

I_D – the index of relevance of the R&D cost, calculated using a statistical database for similar R&D;

I_S – index of relevance of the R&D cost justification, obtained by analyzing the stages of R&D.

All indexes are evaluated on a single scale of points, allowing fractional values of points.

As a result of calculating private indices, it is possible to obtain an aggregated general indicator of the relevance of justifying the R&D cost for the development of helicopter equipment (units, machines, systems) according to the following formula:

$$I = 0{,}65 \cdot I_D + 0{,}35 \cdot I_S.$$

It can be seen from this formula that the R&D cost relevance indicator has a greater contribution to the aggregated indicator, since when justifying the cost it is important how the offered cost is relevant to the cost calculated using the R&D statistical base.

The result of this methodology provides an average score for the relevance of the cost justification in the range from 0 to 5.

The use of several stages of the methodology, the general index of the relevance of costs required to perform R&D is calculated using special averaging formulas.

3.2 METHODS OF OPTIMAL PLANNING OF ORGANIZATIONAL AND TECHNICAL ACTIONS IN PRODUCTION UNDER RESOURCE CONSTRAINTS

Implementation of innovative projects of high-tech production is carried out through various organizational and technical actions, resulting in economic and technological effect. However, these measures affect the outcome with varying degrees of effectiveness. Therefore, we will consider the following economic and mathematical model of optimal planning of organizational and technical actions for the implementation of innovative projects under resource constraints.

We will suppose that for the implementation of an innovative project it is necessary to carry out M different organizational and technical activities. We compare each event numbered m with a number denoted by X_m. We will interpret this number as follows. If this action is not carried out, then we assume

$$X_m = 0.$$

If the action is completed completely, then

$$X_m = 1.$$

Besides, we will consider the values

$$0 < X_m < 1,$$

that we will interpret as a partial implementation of this action X_m.

Since real organizational and technical actions cannot always be partially implemented, we will consider the values of X_m with the following constrains

$$X_m \geq v_m,$$

where the numbers $v_m \in [0, 1]$ show the minimum part of the possible performance of the action m. If $v_m = 1$, then this means

85

that activity m must be performed completely, or it will not be performed at all. We will assume that N different resources are allocated for the implementation of organizational and technical actions. We introduce numerical values

$$B_m \geq 0,$$

that express the volume of each resource.

Since different actions consume different types of resources, we introduce the following values:

$$a_{ij} \geq 0,$$

that have the following meaning. The value a_{ij} shows how much of the i-th resource is consumed by the j-th event.

Planning of organizational and technical actions for the implementation of innovative projects under resource constraints lies in choosing such X_m values that the following inequalities are met

$$a_{11} \cdot X_1 + a_{12} \cdot X_2 + \ldots + a_{1M} \cdot X_M \leq B_1,$$
$$a_{21} \cdot X_1 + a_{22} \cdot X_2 + \ldots + a_{2M} \cdot X_M \leq B_2,$$
$$\ldots \tag{3.1}$$
$$a_{N1} \cdot X_1 + a_{N2} \cdot X_2 + \ldots + a_{NM} \cdot X_M \leq B_N.$$

These inequalities reflect the requirements related to general resource constraints.

In addition, the following restrictions must be met

$$v_1 \leq X_1 \leq 1 \text{ or } X_1 = 0,$$
$$v_2 \leq X_2 \leq 1 \text{ or } X_2 = 0,$$
$$\ldots \tag{3.2}$$
$$v_M \leq X_M \leq 1 \text{ or } X_M = 0.$$

These restrictions reflect the fact that all actions must be completed with a proportion not less than the allowed one, but it is possible that some activities will not be completed at all. Through

$$X = (X_1, X_2, \ldots, X_M)^T$$

we will denote a column vector of X_m values and will call this vector the plan of organizational and technical actions for the implementation of the considered innovative project. We will consider the plan X to be valid if the values X_m included in X simultaneously satisfy the conditions (3.1) and (3.2). The set of all valid plans of organizational and technical actions is denoted by U^X

$$X \in U^X.$$

Suppose that the set of admissible plans for organizational and technical measures is non-empty

$$U^X \neq \emptyset.$$

In this formalism, the problem of optimal planning is to choose such a valid plan of organizational and technical actions $X \in U^X$, which will be most effective to ensure the considered innovative project. For this, it is necessary to introduce a criterion that will evaluate this efficiency.

As such a criterion, consider the numerical functional F, given on the set of valid plans of organizational and technical actions

$$F : U^X \rightarrow R.$$

Based on the economic sense, we will assume that the following condition for the functional efficiency of organizational and technical measures plans takes place:

$$F(X) \geq 0, \text{ for all } X \in U^X.$$

In addition to notation F (X), you can use the equivalent notation

$$F(X) = F(X_1, X_2, ..., X_M).$$

In addition to this condition, this functional must meet the following conditions

$$F(X_1, ..., X_m, ..., X_M) \leq F(X_1, ..., X'_m, ..., X_M),$$

$$\text{if } X_m \leq X'_m.$$

This condition means that if we implement an action to a greater extent while maintaining the remaining actions, the effectiveness of this plan of organizational and technical actions will not be worse than the original plan.

As a typical example of the functional of the efficiency of organizational and technical actions for the implementation of an innovative project, we can offer a linear functional, which looks like this:

$$F(X_1, X_2, ..., X_M) = C_1 \cdot X_1 + C_2 \cdot X_2 + ... + C_M \cdot X_M,$$

where the coefficients $C_m > 0$ show the economic meaning of the impact of the m-th event on the efficiency of the implementation of an innovation project.

Another form of functional for assessing the efficiency of organizational and technical actions is the multiplicative form

$$F(X_1, X_2, ..., X_M) = K \cdot (X_1)^{\gamma_1} \cdot (X_2)^{\gamma_2} \cdot ... \cdot (X_M)^{\gamma_M},$$

where $\gamma_m > 0$. It is usually assumed that the condition

$$0 < \gamma_m \leq 1, m = 1, 2, ..., M.$$

In multiplicative form, it is assumed that all valid X plans satisfy the conditions

$$X_m > 0,$$

since otherwise the functional F will be equal to zero. This case describes a situation where all organizational and technical actions are necessary for the implementation of the considered innovative project.

The general scheme for the optimal planning of these organizational and technical action consists in solving the following extreme problem

$$F(X_1, X_2, ..., X_M) \to \max_{X1, X2, ..., XM.} \tag{3.3}$$

We will assume that a valid plan of organizational and technical actions X* is optimal if the following condition is met:

$$F(X^*) = \max\{F(X): X \in U^X\}.$$

It can be shown that the set U^X is closed and restricted, therefore problem (3.3) always has a solution (not necessarily one). For the actual calculation of the optimal plan, various well-known methods of mathematical programming and optimization theory can be applied.

Methods based on selecting a set of valid plans.

When building a model for optimal planning of organizational and technical actions for the implementation of innovative projects under resource constraints, one of the main problems is the choice of a set of valid plans. During the initial planning, sometimes conditions (3.1) and (3.2) will lead to the fact that the set of U^X valid plans turns out to be empty, which is linked to the fact that the allocated resources are insufficient to implement all the necessary organizational and technical actions. In this situation, the following methods should be followed:

elimination of non-critical actions;

reducing resource consumption by individual actions;

reallocation of resources;

Let us look at these methods closely.

The elimination of non-critical actions when planning innovative projects allows reducing the necessary resources. Mathematically, this method assumes that planning is valid, in which there will be zero values for some actions

$$X_m = 0, m \in M',$$

where M' is a subset of not critical indices of organizational and technical actions within an innovative project.

The following approach can be used to define this set:

1. allow $X_m = 0$ for all indices. In this case, the set of valid plans of organizational and technical actions should become non-empty;
2. using this set of U^X, find the optimal plan by solving problem (3.3);
3. analyzing the optimal plan of measures X^*, find measures with the minimum values of X^*_m;
4. among the found events with minimum values of X^*_m, allow the presence of non-critical actions.

Another approach for detecting non-critical organizational and technical actions is the method of analyzing the functional evaluation of the efficiency of plans.

To do this, it is necessary to calculate the partial derivatives of the functional evaluation of the efficiency of organizational and technical actions at an innovative project

$$F = \frac{\partial F(1, 1, \ldots, 1)}{\partial X_m}.$$

In this formula, partial derivatives are calculated for argument values equal to 1.

Analyzing the values of F_1, F_2, ..., F_M, we need to find those that have the smallest values. If F_m has a small value, then the final functional for assessing the efficiency of organizational and technical actions is weakly dependent on the value of Xm, so action X_m can be excluded from the mandatory measures.

Reducing the resource consumption of individual activities allows reducing the overall resource requirements, which makes it possible to expand the set of valid action plans. In the presented mathematical model, this means that we can reduce the numerical values of the coefficients a_{ij}. Since the values of X_m are non-negative, the set of valid plans should increase as the values of a_{ij} decrease. Indeed, for coefficients a'_{ij} such as

$$a'_{ij} \leq a_{ij},$$

always have

$$a'_{11} \cdot X_1 + a'_{12} \cdot X_2 + \ldots + a'_{1M} \cdot X_M \leq a_{11} \cdot X_1 + a_{12} \cdot X_2 + \ldots + a_{1M} \cdot X_M \leq B_1,$$

$$a'_{21} \cdot X_1 + a'_{22} \cdot X_2 + \ldots + a'_{2M} \cdot X_M \leq a_{21} \cdot X_1 + a_{22} \cdot X_2 + \ldots + a_{2M} \cdot X_M \leq B_2,$$

$$\ldots$$

$$a'_{N1} \cdot X_1 + a'_{N2} \cdot X_2 + \ldots + a'_{NM} \cdot X_M \leq a_{N1} \cdot X_1 + a_{N2} \cdot X_2 + \ldots + a_{NM} \cdot X_M \leq B_N.$$

Therefore, the set of valid plans for organizational and technical actions in the implementation of innovative projects, built using the coefficients a'_{ij}, which we denote by U'^M, satisfies the following inclusion

$$U^M \subset U'^M.$$

Of course, reducing resource consumption by each action requires a certain change in the organizational and technical actions themselves, but this work will significantly increase the economic efficiency of actions to support innovative projects in the future.

Redistribution of resources is another method that allows to optimize the set of valid plans for organizational and technical actions in the implementation of an innovative project. Since various activities require unequal resources, the initial resource constraints can create a "bottleneck" situation, when one of the resources becomes scarce, which does not fully implement all the necessary organizational and technical measures.

Resource constraints arise from financial constraints on the actions to support an innovative project, so there is a possibility of reallocation of resources while maintaining the total amount of financial constraints. To describe this process, we introduce the following notation

$$\varphi_n = \varphi(B_n), \ n = 1, 2, \ldots, N,$$

which mean the cost of the n -th resource in the volume B_n.

Thus, the total cost of all considered resources can be expressed by the following formula:

$$\Phi = \varphi_1 + \varphi_2 + \ldots + \varphi_N.$$

This value depends on the set of resources

$$\Phi = \Phi \, (B_1, B_2, \ldots, B_N).$$

We will assume that the value is fixed and cannot be increased, but the B_n values can be changed.

Under the operation of reallocating resources, we will consider the represenatations

$$\psi_1(B_1) = B'_1,$$

$$\psi_2(B_2) = B'_2,$$

$$\ldots$$

$$\psi_N(B_N) = B'_N.$$

The following resource constraints are met

$$\Phi' = \Phi(\psi_1(B_1), \psi_2(B_2), \ldots \psi_N(BN)) \leq \Phi(B1, B2, \ldots, B_N) = \Phi.$$

These constraints mean that in the process of reallocation of resources, the total amount of financial constraints does not increase, therefore, such a reallocation can be considered economically feasible.

At the same time, it is necessary to consider separately the issue of resources that are not always financially equivalent. In this model, we will analyze only material resources, which can always be expressed by their financial value.

Questions related to random factors in the optimal planning of organizational and technical actions.

Since innovative processes in knowledge-based organizations are always accompanied by certain economic risks, in the optimal planning of organizational and technical actions it is necessary to take into account random factors that can have a significant impact on the result of these activities. These factors are particularly important in the context of resource constraints.

Let us review the options for the negative impact of random factors:

- factors that reduce the economic effect of organizational and technical measures;
- factors that increase the need for resources;
- factors that reduce available resources.

Random factors can lead to decrease in the economic effect of organizational and technical actions in the implementation of innovative projects. According to the developed economic and mathematical model, the functional for assessing the efficiency of these actions should be modified as follows:

$$F(X) = F(X_1, X_2, \ldots, X_M) - \xi_1(X_1) - \xi_2(X_2) - \ldots - \xi_M(X_M).$$

Here $\xi_m(\omega, X_m)$ are random variables that satisfy the conditions with probability up to one

$$\xi_m(\omega, \cdot) \geq 0, \ m = 1, 2, \ldots, M.$$

In these random variables, X_m are the parameters that determine the distributions of random variables.

We will assume that the following conditions are satisfied for all m, which we formulate in terms of the first two moments (expectation and variance):

$$E[\xi_m(\omega, X_m)] \leq E[\xi_m(\omega, X'_m)]$$

and the condition

$$D[\xi_m(\omega, X_m)] \leq D[\xi_m(\omega, X'_m)],$$

$$\text{if } X_m \leq X'_m.$$

These conditions mean that the mathematical expectation and variance of the corresponding random variables increase with the growth of the value of the implementation of an organizational and technical action.

Random factors that increase the need for resources are the result of the consumption of a large number of resources during organizational and technical activities in the implementation of innovative projects than previously planned, since innovative processes are always accompanied by the development and implementation of new technologies.

From the point of view of the mathematical model of optimal planning of organizational and technical actions in the implementation of innovative projects, these factors are reduced to the fact that the coefficients a_{ij} must be modified considering these factors:

$$a_{ij} \to a_{ij} + \chi_{ij}(\omega), i = 1, 2, \ldots, M; j = 1, 2, \ldots, N.$$

Here we review random variables $\chi_{ij}(\omega)$, that satisfy the following conditions:

$$\chi_{ij}(\omega) \geq 0.$$

Thus, the considered random factors can really increase the need for resources in the implementation of organizational and technical measures.

Random factors that reduce available resources arise from the long-term implementation of innovative processes, so various economic factors can lead to a reduction in the financing of innovative processes and, accordingly, to a decrease in available resources for organizational and technical actions.

According to the considered mathematical model, to address these factors it is necessary to change the values of B_n as follows:

$$B_n \to B_n + \zeta_n(\omega),$$

where the random variables $\zeta_n(\omega)$ meet the following conditions

$$\zeta_n(\omega) \leq 0, \, n = 1, 2, \ldots, N.$$

Since the values of random variables are usually independent of the selected plans, to predict their values, it is necessary to consider various external random factors and risks.

To address these factors for optimal planning of organizational and technical actions in the implementation of innovative processes under resource constraints, it is necessary to consider the possibility of reducing the values of B_n when solving the extreme problem (3.3). For this, it is feasible to deliberately reduce the set of admissible plans of organizational and technical measures based on constraints (3.1), where some values of B_n will be reduced given possible risks. Of course, this may lead to less effective plans being selected, but it compensates for the possible risks of resource shortages when implementing critical organizational and technical actions.

Eventually using the example of an economic and mathematical model, we have clearly demonstrated how the results of intelligent monitoring contribute to improving the competitiveness of high-tech production (for example, helicopter engineering). This model describes a set of valid plans of organizational and technical actions for the implementation of innovative projects under resource constraints. Through the module, a general principle for building optimal plans is formulated, which in turn create breakthrough technologies and ensure the success of product sales on the market.

The efficiency of intelligent monitoring of high-tech production (in particular, helicopter engineering) is justified by solving a wide range of problems:

1. collection and storage of parameters and data of intelligent monitoring;
2. analysis of changes in key parameters of equipment, process and production in general;

3. analysis of the effective use of complexes in production technology;
4. optimization of process chains by numerous criteria;
5. optimization of maintenance of production complexes.

The above methods can undoubtedly be used in solving production optimization problems. In order to develop effective consistent recommendations based on the intelligent monitoring results, it is important to correctly determine the conditionally constant values of parameters during the simulation period.

3.3 COST OPTIMIZATION AT VARIOUS STAGES OF THE LIFE CYCLE OF INNOVATIVE PROJECTS FOR THE CREATION OF HIGH-TECH PRODUCTS

When creating high-tech products, innovative projects are implemented at various stages of the life cycle of these projects. Moreover, each stage requires certain expenditure (financial, material, labor), as a result of which the total cost of implementing innovative projects is formed. The specificity of such projects is that expenditure at different stages of the life cycle can not always be planned in advance and, in addition, the variation of expenditure at some stages of the life cycle leads to a change in the expenditure at other stages. The reason for this is that innovative projects are connected with exploratory scientific research and new tasks that need to be addressed in their implementation.

Thus, there is an urgent problem of developing a methodology for optimizing costs at various stages of the life cycle of innovative projects for the creation of high-tech products.

Cost optimization for these projects is important, since it can significantly reduce the cost of products and increase the competitiveness of products. When planning actions to optimize the costs of implementing innovative projects, it is necessary not only to apply standard methods of cost reduction, but also to use optimization at all stages of the life cycle, since the specificity of innovative production is that costs at different stages of the life cycle are interdependent. The method of optimizing such costs will significantly reduce the cost of developing and implementing innovative technologies in the knowledge-based industry by optimizing each stage of the life cycle of innovative projects, given the main goal of reducing the cost of developing and implementing innovative technologies.

Consider a certain innovative project for the creation of high-tech products, consisting of M different stages. We will assume that these stages are numbered in the order of their implementation in an innovation project

$$T_1, T_2, \ldots, T_M, \tag{3.4}$$

where the m-th stage of the innovation project is denoted by T_m.

Note that in many cases, the stages of the life cycle of innovative projects can be carried out in parallel and then they cannot always be ordered by time. In this case, the numbering (3.4) can be done at random for those stages that are carried out simultaneously. However, the methodology assumes that the last stage of the T_M life cycle should be the final stage of an innovative project.

Using S_m, we will denote the funding of the m-th stage of the life cycle of an innovative project. Accordingly, through

$$S = S_1 + S_2 + \ldots + S_M$$

we denote the total amount of funding for the innovation project. The financing of each stage of an innovative project will be evaluated using the efficiency functional of the life cycle stage. We represent these functionals as

$$F_m = F_m(x), \quad m = 1, 2, \ldots, M,$$

where x is the corresponding funding for the life cycle stage.
The F_m functions must meet the following conditions:

$$F_m(x) > 0,$$

$$F'(x) \geq 0, \tag{3.5}$$

$$F''(x) \leq 0.$$

The first condition says that the efficiency of each stage should be positive, the second states that with an increase in funding, the efficiency of a life cycle stage of an innovative project should not decrease. Finally, the third condition is that with an increase in funding, the additional increase in efficiency should decrease, which corresponds to the fundamental economic law of marginal utility.

The choice of these functionals for assessing the efficiency of financing the stages of innovative projects is a separate task, but in many practical cases these functions can be chosen in the form of power functions:

$$F_m(x) = A_m \cdot x^{\alpha_m},$$

where $A_m > 0$ is the dimension coefficient; α_m is the exponent that must satisfy the conditions

$$0 < \alpha_m \leq 1.$$

Functions of this type satisfy all conditions (3.5). An approximate view of this function is shown in Fig.3.1.

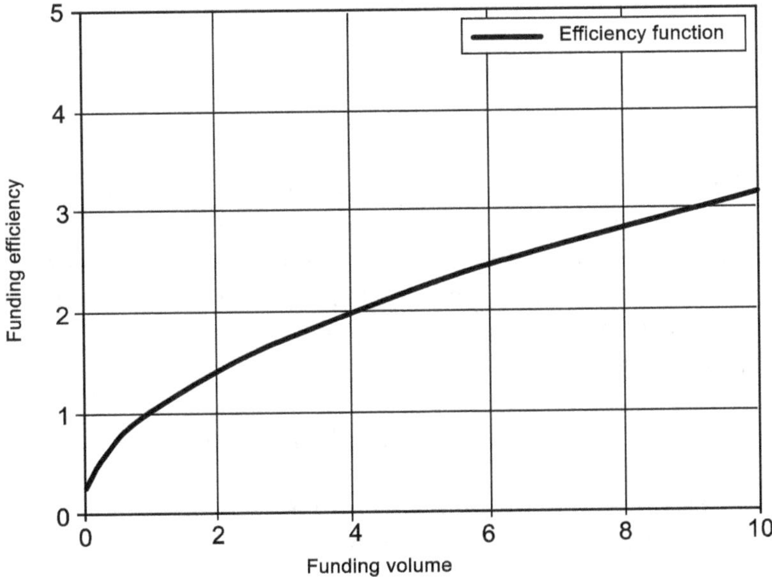

Figure 3.1. Approximate model of the efficiency function depending on the funding volume

The overall assessment of the efficiency of the distribution of funding for the stages of the life cycle of innovative projects is expressed by the following formula

$$F(S_1, S_2, ..., S_M) = F_1(S_1) + F_2(S_2) + ... + F_M(S_M). \qquad (3.6)$$

The task of the method of cost optimization at various stages of the life cycle of innovative projects for the creation of high-tech products is to find such a distribution of funding for the stages of the life cycle, when the overall efficiency will be sufficient with minimal total funding. These needs to take into account the restrictions on the minimum financing of each stage of the life cycle, i.e. the distribution of funding must meet the condition

$$S_m \geq MS_m, m = 1, 2, ..., M, \qquad (3.7)$$

where MS_m is the minimum funding for the m-th stage.

Also, when considering the minimum efficiency of an innovative project, the condition must be met

$$F(S_1, S_2, ..., S_M) \geq MF, \qquad (3.8)$$

where MF is the minimum overall efficiency of the innovation project.

The purpose of the cost optimization methodology at various stages is to reduce the total costs of the implementation and introduction of innovative technologies when creating high-tech products at knowledge-based enterprises, which should increase the competitiveness of high-tech products. This method solves the optimization problems:

- expenditure at each stage of the life cycle of an innovative project;
- total costs of an innovative project while maintaining the required efficiency of the innovative project.

Using this methodology, the following indicators are determined:

- optimal funding of each stage of the life cycle of an innovative project;
- overall optimal funding of an innovative project.

The following parameters are used as the initial data for the cost optimization methodology at various stages of the life cycle of an innovative project to create high-tech products (Table 3.3).

Table 3.3. Initial data for the cost optimization methodology
at various stages of the life cycle of an innovative project

№	Parameter name	Notation	Unit of measurement
1	List of stages in the life cycle of an innovative project	T_m	–
2	Coefficients of the efficiency functions of funding the stages of the life cycle of an innovative project	A_m	Dimensionless
3	The degree of efficiency functions of funding the stages of the life cycle of an innovative project	α_m	Dimensionless
4	Minimum level of funding for each stage of the innovation project life cycle	MS_m	RUB.
5	Minimum efficiency of an innovative project	MF	Dimensionless
6	Step of changing the total amount of financing for an innovation project	ΔS	RUB.

The cost optimization methodology algorithm is implemented in several steps.

Step 1. Presentation of an innovative project by a sequence of stages of the life cycle.

At this point, it is necessary to form the stages of the life cycle of an innovative project, i.e. create a list of T_m stages in such a way that this sequence of stages is arranged in chronological order.

Step 2. Selecting the amount of funding for the entire project.

We set the value of S according to preliminary estimates of the level of funding of an innovative project. At the same step, we take the value of a logic variable

Flag = False.

Step 3. Solving the optimization problem for a given S.

We solve the following optimization problem:

$$F_1(S_1) + F_2(S_2) + \ldots + F_M(S_M) \to \max$$

Upon the conditions

$$S = S_1 + S_2 + \ldots + S_M,$$

$$S_m \geq MS_m, \ m = 1, 2, \ldots, M.$$

In another algorithm, we will look at the way for solving this problem.

Step 4. Adjusting the S value.

As a result of step 3, we have an optimal distribution of funding at various stages of the life cycle of an innovative project at a given value of S. In this case, if the condition is met

$$F(S_1, S_2, \ldots, S_M) < MF,$$

then if the Flag variable is False, then it is necessary to increase the value of S

$$S := S + \Delta S.$$

After that, we return to step 3.

If the Flag variable is True, then we accept

$$S := S + \Delta S$$

and finish the algorithm.

If the condition is

$$F(S_1, S_2, \ldots, S_M) \geq MF,$$

then it is necessary to reduce the value of S

$$S := S - \Delta S$$

and accept

$$Flag = True.$$

After that, we need to return to step 3.

As a result of this algorithm, we obtain the minimum and optimal cost values at various stages of the life cycle of innovative projects to create high-tech products.

The above algorithm contains step 3, which solves the optimization problem. This problem is a variant of the dynamic programming problem.

We describe an algorithm for solving this optimization problem.

Step 1.

Let us denote by $U_M(R)$ the value

$$U_M(R) = F_M(R).$$

Step 2.

Then for all the steps $m = M - 1, M - 2, \ldots, 2$.

Then

$$F_m(x) + U_{m+1}(R - x) \to \max$$

$$\text{if } x \leq R.$$

The solution of these problems is denoted by $Um(R)$. $U_m(R)$.

Step 3.

Let us assume that

$$U_1(R) = \max\{F_1(x) + U_2(R - x)\} \qquad (3.9)$$

$$\text{if } x < R.$$

Step 4.

The value x at which the maximum (3.9) is realized, if $R = S$, is denoted by S_1.

Step 5.

For each $m = 2, 3, \ldots, M - 1$ we find S_m as follows. The quantity S_m is the x at which the maximum is

$$\max\{F_m(x) + U_{m+1}((S - S_1 - S_2 - \ldots - S_{m-1}) - x)\}.$$

We define S_M as follows

$$S_M = S - S_1 - S_2 - \ldots - S_{m-1}.$$

Thus, we get a set of values $S_1, S_2, \ldots, S_M$, which are the solution to the optimization problem.

The result of using the cost optimization methodology is the optimal financing of the entire innovation project S and the optimal distribution of $S_1, S_2, \ldots, S_M$ costs at each stage of the life cycle of the innovation project.

An example of using the methodology

Let us give a model example of using the cost optimization methodology at various stages of the life cycle of an innovative project to create high-tech products.

We will consider an innovative process for creating a rotor blade using new composite materials.

We will assume that the life cycle of this innovative process includes five stages:

1. preparatory stage, product sketch development;
2. stage of conducting search and research works;
3. product design and production stage;
4. testing and adjusting the image of the product;
5. introduction of the new product into production.

Let us define the model values of the initial data.

We will use the following values as coefficients of the performance evaluation functions for each stage of the innovation project:

$A_1 = 0,6,$
$A_2 = 1,$
$A_3 = 1,$
$A_4 = 0,9,$
$A_5 = 0,7,$
$\alpha_1 = 0,5,$
$\alpha_2 = 0,45,$
$\alpha_3 = 0,5,$
$\alpha_4 = 0,4,$
$\alpha_5 = 0,5.$

We will set the minimum level of funding for each stage of the life cycle of the considered innovative project:

$MS_1 = 10$ million rubles;
$MS_2 = 17$ million rubles;
$MS_3 = 27$ million rubles;
$MS_4 = 9$ million rubles;
$MS_5 = 14$ million rubles.

The overall efficiency of the innovation project is set as follows:

MF = 45 conventional units.

The step of changing the total amount of funding for a project amounts to

ΔS = 10 million rubles.

Let us start the calculation with the initial amount of funding for the entire project, equal to S = 150 million rubles.

Calculations, according to the proposed methodology, show that in this case we have the following optimal distribution of costs over the stages of the life cycle of an innovative project (funding):

stage 1 – 19.5 million rubles;
stage 2 – 31.32 million rubles;
stage 3 – 55.5408 million rubles;
stage 4 – 16.5829 million rubles;
stage 5 – 27.0563 million rubles.

Hence, the efficiency of financing the stages of the life cycle of an innovative project in conventional units will be:

stage 1 – 21,2219;
stage 2 – 18,5724;
stage 3 – 13,8613;
stage 4 – 6,40872;
stage 5 – 3,6411.

The overall efficiency of project financing amounts to 63,7 conventional units.

In this case, according to the methodology, it is necessary to reduce the total funding

$$S := S - \Delta S.$$

Let us continue the calculations with the total volume of financing S = 140 million rubles.

Then funding by stages of the life cycle will be equal to:

stage 1 – 18.2 million rubles;
stage 2 – 29.232 million rubles;
stage 3 – 51.8381 million rubles;
stage 4 – 15.4774 million rubles;
stage 5 – 25.2526 million rubles

Funding efficiency by stage is equal to:

stage 1 – 20.5365;
stage 2 – 17.9768;
stage 3 – 13.4098;
stage 4 – 6.20992;
stage 5 – 3.51763.

The overall efficiency of project financing amounts to 61.7 conventional units.

We reduce the total funding to RUB 130 million. After calculations, we get financing by stages:

stage 1 – 16.9 million rubles;
stage 2 – 27.144 million rubles;
stage 3 – 48.1354 million rubles;
stage 4 – 14.3718 million rubles;
stage 5 – 23.4488 million rubles.

Funding efficiency by stage is equal to:

stage 1 – 19,8251;
stage 2 – 17,3586;
stage 3 – 12.9413;
stage 4 – 6,00333;
stage 5 – 3,38968.

The overall efficiency of project financing amounts to 59.5 conventional units.

We consistently reduce the total funding to 120 million rubles. After the calculations, we receive funding by stages:

stage 1 – 15.6 million rubles;
stage 2 – 26.1 million rubles;
stage 3 – 43.848 million rubles;
stage 4 – 13.4363 million rubles;
stage 5 – 21.0157 million rubles.

Funding efficiency by stage is equal to:

stage 1 – 19.0848;
stage 2 – 16.715;
stage 3 – 12.375;
stage 4 – 5.75322;
stage 5 – 3.209.

The overall efficiency of project financing amounts to 57.1 conventional units.

We reduce the total funding to 110 million rubles. After calculations, we get funding for the stages:

stage 1 – 14.3 million rubles;
stage 2 – 23.925 million rubles;
stage 3 – 39.4763 million rubles;
stage 4 – RUB 12.5965 million;
stage 5 – 19.7022 million rubles.

Funding efficiency by stage is equal to:

stage 1 – 18.3118;
stage 2 – 16.0428;
stage 3 – 11.8695;
stage 4 – 5.58648;
stage 5 – 3.1071.

The overall efficiency of the financing of the project amounts to 54,9 conventional units.

We reduce the total funding to 100 million rubles. After the calculations, we get funding for the stages:

stage 1 – 13 million rubles;
stage 2 – 21.75 million rubles;
stage 3 – 35.8875 million rubles;

stage 4 – 11.4514 million rubles;
stage 5 – 17.9111 million rubles.

The funding efficiency of the stages is equal to:

stage 1 – 17,5012;
stage 2 – 15,3379;
stage 3 – 11,3398;
stage 4 – 5,34914;
stage 5 – 2.96251.

The overall efficiency of project financing amounts to 52.5 conventional units.

We reduce the total funding to 90 million rubles. After calculations, we get funding for the stages:

stage 1 – 11.7 million rubles;
stage 2 – 19.575 million rubles;
stage 3 – 32.2988 million rubles;
stage 4 – 10.5705 million rubles;
stage 5 – 15.8557 million rubles.

The funding efficiency by stage is equal to:

stage 1 – 16.6473;
stage 2 – 14.595;
stage 3 – 10.782;
stage 4 – 5.09878;
stage 5 – 2.78735.

The overall project financing efficiency amounts to 49.9 conventional units.

We reduce the total funding to 80 million rubles. After calculations, we get funding for the stages:

stage 1 – 10.4 million rubles;
stage 2-17.4 million rubles;
stage 3 – 28.71 million rubles;
stage 4 – 9.396 million rubles;
stage 5 – 14.094 million rubles.

The funding efficiency by stage will be equal to:

stage 1 – 15.7423;
stage 2 – 13.8073;
stage 3 – 10.1912;
stage 4 – 4.83299;
stage 5 – 2.62794.

The overall project financing efficiency amounts to 47.2 conventional units.

We reduce the total funding to 70 million rubles. After the calculations, we get funding for the stages:

stage 1 – 9.1 million rubles;
stage 2-15.225 million rubles;
stage 3 – 25.112 million rubles;
stage 4 – 8.42704 million rubles;
stage 5-12.1267 million rubles.

The funding efficiency by stage will be equal to:

stage 1 – 14.7761;
stage 2 – 12.9661;
stage 3 – 9,56087;
stage 4 – 4,54876;
stage 5 – 2,43764.

The overall efficiency of project financing amounts to 44.3 conventional units. The obtained efficiency is below the minimum level of efficiency MF = 45 conventional units, therefore, as a result of applying the methodology, it is necessary to use the total funding S equal to 80 million rubles, distributing it by stages of the life cycle as follows:

stage 1 – 10.4 million rubles;
stage 2 – 17.4 million rubles;
stage 3 – 28.71 million rubles;
stage 4 – 9.396 million rubles;
stage 5 – 14.094 million rubles.

Herewith, the resulting distribution meets the specified minimum funding conditions for each stage of the life cycle.

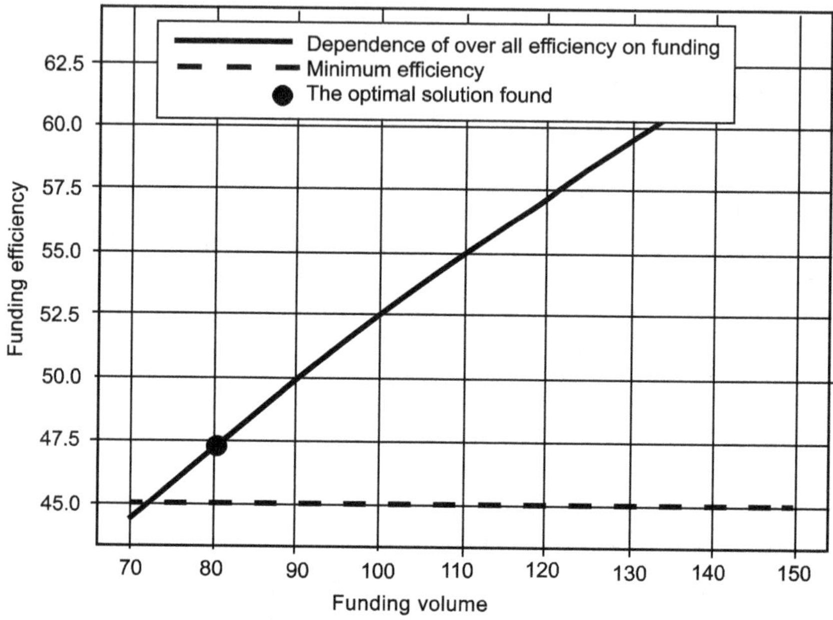

Figure 3.2. Calculation results using the cost optimization method at various stages

The results of the calculations are shown in Fig. 3.2.

The proposed method of cost optimization at various stages of the life cycle of innovative projects for the creation of high-tech products makes it possible to determine the optimal level of financing for an innovative project to create promising products, which provides a given level of project financing efficiency. Moreover, the algorithms of the method allow to find the optimal distribution of funding for various stages of the life cycle of an innovative project.

CHAPTER 4. OPTIMAL COST FORMATION OF HIGH-TECH PRODUCTS USING THE ADVANTAGES OF DIGITAL TECHNOLOGIES

4.1. ASSESSMENT OF THE IMPACT OF MODERN DIGITAL TECHNOLOGIES ON THE PRODUCT VALUE CHAIN

The Russian economy is production and industrial oriented, in particular focused on mining and processing industry, which use the achievements of the third and fourth technological modes allowing the transition to the digital economy. The main task of digital transformation is to increase the growth of added value of production, which should occur through the integration of enterprises that combine all stages of the product life cycle.

The level of Russia's technological potential, fundamental and applied scientific achievements, and the integration of production technologies into a single complex using intelligent technologies are the most valuable competitive advantages of the country in the context of digital transformation.

Let us review an economic and mathematical model that can be used to create a methodology for assessing the impact of digital technologies on the product value chain and transforming it into a tool for increasing competitiveness. This chain involves a large number of factors that ultimately increase the total cost of products, so to maintain competitiveness, organizations must take continuous measures to reduce (limit) the cost of products at all stages of the value chain. According to modern approaches to the digital economy, the most important measure for optimizing this chain is the use of digital technologies.

Digital technologies in manufacturing include a large number of different digital and informational tools. These technologies can be grouped as follows:

- digital product records;
- three-dimensional model of components and products;
- digital description of technological and production processes;
- additive technologies (3D printing);
- internet of things and robotization;
- machine learning and artificial intelligence.

All these technologies can be used to optimize the value chain of high-tech products. In doing so, the impact on the value chain of products is possible either in terms of optimizing the production process, or in terms of optimizing the management decisions that affect the cost of high-tech products.

Let us examine the impact of these groups of digital technologies on the product value chain.

Digital records of a product allow getting accurate and up-to-date information about the products being generated. Due to the specifics of digital technologies, product descriptions can be scalable and cover the entire production chain. This opportunity can be used to make more accurate management decisions in the production process, which will reduce the cost of some stages of production. Moreover digital product records enable precise control of production at all stages, which reduces the risk of additional expenditure.

The current economic situation requires significant optimization of all support production processes related to documentation and information support of production processes. The use of digital product records based on modern information and analytical systems and ERP systems can significantly reduce the expenditure on documenting and information support of production processes of knowledge-based products.

The use of electronic management significantly reduces production overhead costs. In particular, replacing the routine work of creating documentation and information support with modern automated digital technologies offers major opportunities for optimizing support personnel in production. Digital technologies are not only cost-effective, but also allow performing traditional human activities with greater accuracy and efficiency.

Digital processes for documenting manufacturing processes can have a major impact on the entire value chain of knowledge-based enterprises. In addition, digital records in production increase the overall culture of production, and consequently increase the value of the enterprise itself, making it more competitive.

The three-dimensional model of the product component allows presenting the project of the finished product and all its main components not only in the form of traditional drawings, but also in a visual form of three-dimensional design. The model not only gives a three-dimensional image of the product, but also allows carrying out three-dimensional modeling of the product and its components for accurate design of the technical characteristics of the product.

Moreover, the use of three-dimensional digital models of technological products provides geometric optimization of a product's image, which can significantly increase competitiveness of a number of high-tech products.

The influence of the 3D model of the component and product on the product value chain is carried out through possible optimization in product design. In particular, with a three-dimensional digital model, it is possible to carry out optimization procedures aimed at reducing consumed materials, reducing the cost of product components, creation processes, as well as product assembly. This optimization can have a significant impact on the entire chain.

Digital description of technological and production processes is a great opportunity to optimize the value chain of products, since the closed information cycle of production allows for global optimization of the entire production process. Moreover, the digital format of the representation of technological and production processes provides the opportunity to formulate optimization and extremal mathematical problems and build effective economic and mathematical models for solving the problem of cost optimization in designing and developing products.

The availability of a digital description of technological and production processes makes the production of high-tech products scalable and mobile, since it is less dependent on specific design bureaus, pilot and serial production. The availability of a digital description of technological processes enables the mobility of production processes in a shorter time and at the lowest cost. Also, a digital description of technological and production processes implies precise control over the production processes of high-tech products, which reduces the economic risks related to the failure to achieve the specified characteristics of products, as well as the risk of excess funding of certain production processes.

In modern production of high-tech products, construction materials tend to play an essential role, which are becoming more expensive and more difficult to manufacture. The availability of digital models of technological and production processes allows to analyze the complex structural materials used, as well as to optimize the use of expensive materials in the production of high-tech products, which can have a positive impact on the value chain of products.

Additive technologies (3D printing) provide new production methods that have recently significantly changed many technological and production technologies. Economically, the use of additive technologies will significantly affect the value chain of products. On the one hand, their application requires significant investments and additional expenditure for the organization of production, but, on the other hand, these technologies can, firstly, radically improve the product characteristics, making them highly competitive in the market, and secondly, reduce the cost of many production processes.

These technologies have the greatest impact on the product value chain when building promising, high-tech products, in which various innovative technologies are used. At the same time, 3D printing can significantly reduce the economic risk of excess funding of production processes and the risks related to the failure to achieve the specified parameters for promising products.

The Internet of Things and robotization also affect the value chain of knowledge-based products, as they allow optimizing production processes. Industrial robots have been used in production for a long time, being a traditional means of production not only for high-tech products. In the context of digital technology and digitalization, an increasing role is played by versatile robots that can perform a wide range of functions. In many industries, the use of such robots is very significant, especially in the high-tech production of promising products.

Robotization of production processes has a serious impact on the value chain of products, as it allows, firstly, to reduce the participation of working personnel, which provides a significant reduction in production costs, and secondly, to make many production processes more optimized in terms of time and resources involved.

In particular, the consistent use of robotics in production makes many operations more economical. This applies primarily to saving expensive materials that are used more efficiently when applying digital technologies. Moreover, robotics allows to reduce the time of production processes, which also affects the product value chain.

The use of IoT technologies in manufacturing is a relatively new approach to the organization of production processes. However, it is fully consistent with the general trend of digitalization of production and the transition to the digital economy. The integration of intellectual capacity into the means of production, as well as the design of new intelligent tools of production, can significantly change the production ecosystem, since the integration of a large number of intelligent machines into a single digital network gives an additional production and economic effect. Therefore, the Internet of Things and related digital technologies will have a significant impact on the product value chain.

Machine learning and artificial intelligence are the cornerstones of digital economy technologies. Mostly these technologies provide a significant development of digital technologies in production. Machine learning methods allow using machine and automatic tools to solve tasks that have traditionally been performed by humans.

Machine learning and artificial intelligence in manufacturing can play either a supporting role or a crucial role. In the case that artificial intelligence methods are used as supporting tools, we are talking about the use of machine tools to facilitate the production processes or increase the accuracy of their execution. In this case, machine learning and artificial intelligence methods will have a significant impact on the product value chain, as they reduce labor costs in production processes.

The use of artificial intelligence methods as the main tool allows implementing production processes with maximum efficiency, which will provide some operations without human intervention. In this case, we can expect a significant reduction in production costs, which will effectively affect the product value chain.

It should be noted that the methods of machine learning and artificial intelligence can be used to control the network of other intelligent machines, which enables the design production with minimal participation of human labor resources. This approach is increasingly used in the production of high-tech products and significantly changes the product value chain.

Let us take a closer look at the economic and mathematical model for assessing the impact of digital technologies on the product value chain.

We will imagine a time scale T, during which the product value chain will be formed. We assume that

$$T \subset R, \forall\, t \in T, -\infty < t < \infty.$$

As T, we can consider either the segment $T = [a, b]$, or a discrete set of time intervals

$$T = \{t_0, t_1, \ldots, t_K\}.$$

Usually, the value chain of products is studied on a discrete set, but to formulate a mathematical model, it is convenient to consider a continuous time scale.

By F_T we denote the σ-algebra of subsets of the set T. The set of all subsets can always be chosen as F_T.

To assess the impact of digital technologies on the product value chain, we consider the measure λ given on the σ-algebra F_T:

$$\forall\, A \in F_T, \lambda(A) \geq 0.$$

The economic interpretation of this measure is that for a given measurable set $A \in FT$, a nonnegative number $\lambda(A)$ means an increase in the cost of production over the time interval A. The λ measure will reflect a non-digital product value chain. In this case, the total cost of production is expressed by the Lebesgue integral as λ

$$S = \int_T \lambda(dt), \qquad\qquad (4.1)$$

where the integral is calculated over the entire set T.

In the case when we consider a discrete set T, the integral (4.1) is the sum

$$S = \int_T \lambda(dt) = \sum_T \lambda(t_k),$$

where the summation is carried out over all values of t_k, and the measure concentrated at the points of the set T is analyzed.

Let now various digital technologies are used to manufacture the product in question. In this case, instead of the measure λ, we have some other measure, which we denote by μ. The measure μ must also be defined on the set F_T. Moreover, from natural assumptions we have a measure μ that is absolutely continuous with respect to the measure λ, i.e.

$$\forall\, A \in F_T, \text{ using that } \lambda(A) = 0 \text{ it follows that } \mu(A) = 0.$$

Then, by the famous functional analysis Radon-Nikodim theorem, we obtain that there exists a nonnegative function g(t) given on T and measurable on this set. This function is called the Radon-Nikodym derivative

$$g(t) = d\mu \,/\, d\lambda.$$

The function g(t) is defined by the following relation

$$\mu(A) = \int_A g(t)\, \lambda(dt).$$

The economic meaning of the function g(t) is that it is an assessment of the impact of digital technologies on the product value chain.

If at any point in time $t \in T$ we have the value $g(t) < 1$, then at this stage of the product value chain, digital technologies have led to a decrease in the cost of this product. If on the set $A \subset T$

$$g(t) = 0,\, t \in A \subset T,$$

then on the time interval A it was possible to completely eliminate expenditure by using digital technologies.

The overall assessment of the impact of digital technology on the entire product value chain is calculated using the following formula:

$$\Delta = \int_T \lambda(dt) - \int_T g(t)\, \lambda(dt) = \int_T (1 - g(t))\lambda(dt).$$

The value $\Delta \geq 0$ shows the total savings from the impact of digital technologies on the product value chain. The Δ value is an absolute impact assessment of digital technologies on the product value chain. The relative assessment of the impact of digital technologies is calculated using the formula

$$\Theta = (1 - \int_T \lambda(dt) \,/\, \int_T g(t)\, \lambda(dt)) \cdot 100\ \%.$$

The value Θ shows by what percentage the cost of products is reduced as a result of the impact of digital technologies.

The considered economic and mathematical model quantifies the impact of digital technologies on the product value chain. In addition to the absolute and relative general assessment, the considered model allows to assess the "density" of the impact of digital technologies. In this regard, it is possible to create recommendations for the further introduction of digital technologies at different stages of the product value chain.

Thus, the presented economic and mathematical model can be used to create a *methodology for assessing the impact of digital technologies on the product value chain.*

The modern economy increasingly relies on digital technologies, which are being introduced not only in the management of organizations, but also in in production itself. Since the development, introduction and use of information technologies are costly operations, it is necessary to have quantitative assessments of the impact on the cost of products for their effective use. To solve this urgent problem, a methodology for assessing the impact of digital technologies on the product value chain has been developed. This technique is of particular relevance when applying innovative technologies in the development of promising products, allowing to increase the competitiveness of products.

Digital technologies play an increasing role in knowledge-based industries in the context of the transition to the digital economy. Economically, it is necessary to have tools for quantifying the efficiency of the use of digital technologies, since their introduction and use require large financial and other resources.

For this purpose, an economic and mathematical model is used, based on the mathematical apparatus of the abstract Lebesgue integral and the procedure for differentiating absolutely continuous measures according to the Radon-Nikodym theorem. The use of Radon- Nikodym derivative, results in obtaining a quantitative assessment of the impact of digital technologies on the product value chain.

The objective of the digital impact assessment methodology is to build constructive and quantitative assessments of its impact, as well as to identify the main elements of the value chain that are more affected by digital technologies.

This technique solves the following tasks:

- assess (quantitatively) the impact of digital technologies on the product value chain;
- obtain an absolute and relative assessment of the impact of digital technologies on the product value chain;
- assess the impact of digital technologies on each element of the product value chain.

It allows you to define the following parameters:

- an absolute assessment of the impact of digital technologies on the product value chain;
- relative assessment;
- assessment of the impact of digital technologies on individual elements of the product value chain.

The following parameters are used as initial data to the digital technology impact assessment methodology:

- description of digital technologies used in the production;
- description of the product value chain;
- calculating the value of the product value chain without using digital technologies.

Thus, the methodology for assessing the impact of digital technologies on the product value chain can be represented as a process implemented in several stages.

Stage 1.

Choose the time scale on which the product value chain will be considered. This time scale T can be either continuous or discrete, depending on the problem statement.

Stage 2.

Choose the set FT, which consists of subsets of the set T. The set system F_T must form a σ-algebra. For simplicity, you can always choose the set of all subsets as FT.

Stage 3.

According to the initial data on the product value chain, it is necessary to construct a measure λ on the sets F_T. This measure should reflect an increase in the cost of products over a given time interval.

Stage 4.

Analyzing the impact of digital technologies on each element of the product value chain, build a measure μ, which must be absolutely continuous with respect to the measure λ. The measure μ reflects the impact of digital technologies on individual production processes in the value chain.

Stage 5.

Calculate the Radon-Nicodemus derivative of measure μ with measure λ by the formula

$$g(t) = d\mu\,/\,d\lambda,$$

where the measurable function g(t) is set based on the following relation:

$$\mu(A) = \int_A g(t)\,\lambda(dt).$$

The function g(t) represents the "density" of the impact of digital technologies on the product value chain.

Stage 6.

Determine the absolute assessment of the impact of digital technologies on the entire product value chain using the formula

$$\Delta = \int_T \lambda(dt) - \int_T g(t)\,\lambda(dt) = \int_T (1 - g(t))\lambda(dt).$$

The value $\Delta \geq 0$ reflects the total savings from the impact of digital technologies on the product value chain.

Stage 7.

Calculate the relative assessment impact of digital technologies on the product value chain using the formula

$$\Theta = (1 - \int_T \lambda(dt) / \int_T g(t)\,\lambda(dt)) \cdot 100\,\%.$$

The indicator shows by what percentage the cost of products is reduced as a result of the impact of digital technologies.

In summary, digital technologies have a significant impact on the management of knowledge-based enterprises, as well as on production processes in the context of digital transformation. Many digital technologies have been found to have a large impact on the product value chain.

In this regard, a methodology has been developed to quantify the impact of digital technologies on the product value chain, which makes it possible to absolutely and relatively assess the impact of digital technologies on the value chain of promising products. It is shown that this mathematical model, based on the use of the mathematical apparatus of absolutely continuous measures, can be used as the basis for the methodology for assessing the impact of digital technologies.

4.2. DEVELOPMENT OF DESIGN SOLUTIONS THAT ENSURE OPTIMAL PRODUCTION COSTS

The issue of forming the cost of a product is quite well studied in the scientific literature. Many works are devoted to the issues of economic cost analysis, expenditure calculation, etc. However, the methods of solving this issue do not have integrity and generality; do not provide comparability and the necessary reliability of the results when solving the problem.

Planning the cost of a unit of production is carried out in two opposite directions (Figure 4.1): "down-top", i.e. from the beginning

Formation of product characteristics

Classic approach to expenditure formation

- Formation of product characteristics
- Product development
- Development of the technological process
- Formation of cost estimates

Is the cost reasonable? — NO / YES

- Production
- Periodic cost adjustments

Cost is determined by design and technology

Design for a given cost

- Formation of product characteristics taking into account market and consumer expectations
- Determining the target cost of a product
- Finding a balance between target cost and performance
- Analysis of associated costs
- Supplier analysis
- Design, selection of technical process, formation of alternatives
- Cost forcasts in the early stages of product creation
- Choosing a cost-effective way to achieve the target product characteristics
- Production

The cost price determines the choice of design and technological solutions

Figure: 4.1. Approaches to the production costs formation

of the expenditure formation in the production; in the reverse order "top-down", i.e. starting from the market selling price of the product.

In the first direction, the actual costs per unit of production are formed, in the second one – the marginal costs per unit of production.

Actual expenditures reflect the planned production expenditures, which are calculated depending on the state and culture of a particular production. Marginal expenditures show the level of expenditures, above which production will not give the planned level of profitability.

When determining the cost of production, the calculation is carried out from the formation of actual expenditures to the formation of marginal expenditures.

Based on this cost planning mechanism, there are main methods of cost planning:

- only cost-based planning methods are used to generate actual expenditures;
- market methods of cost planning are used to determine the marginal expenditures.

The sequence of calculating planned expenditures using cost and market methods is diametrically opposite. With cost-based planning methods, the calculation is carried out "top-down" according to the calculation (from material expenditures to general production expenditures). With market methods of cost planning, the calculation is made "down-top" according to the calculation items (from the value of the production cost to material expenditures). Approaches to planning the cost of production will be discussed in detail below.

Among the priority areas of innovative development of the economy, the state has identified digital production technologies, the creation of new materials, development of systems capable of processing large amounts of data, artificial intelligence and machine learning. Such a rapid development of digital technologies carries the risk of the industry's unpreparedness for their effective use and implementation in many business processes. To implement the digital transformation of the industry, it is necessary to have new economic mechanisms for managing projects to create high-tech products. One of these mechanisms should be a mechanism for managing the resource support of projects at various stages of their life cycle, which should be based on modern methods of machine learning and artificial intelligence, allowing to solve the problem of automating many business processes in organizations at all stages of the life cycle (from design to sales) of products and making effective management decisions at various stages of the product life cycle through the use of automated expert systems.

The processes of digital transformation of industry are aimed at increasing the competitiveness of organizations and their products.

For example, mathematical modeling of cost should accompany the entire process of designing a multi-component product (Figure 4.2): design of parts (its cost and labor intensity of manufacturing depend on a number of factors: geometric parameters,

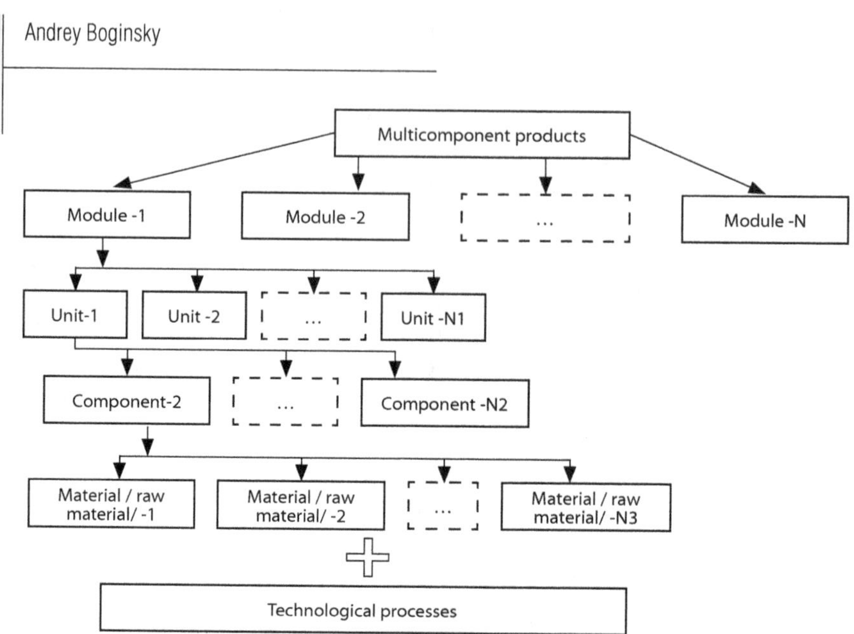

Figure: 4.2. The process of forming the cost of multi-component products

tolerances, fitments, purity of processing, material used, features of technological processes of their manufacture), their interfacing into a unit, module, and final product.

At the part level, using simulation modeling, geometric parameters/ tolerances/fit of the product can be selected, ensuring optimal cost and labor intensity.

Design and economic analysis should be related to each other, and the results of economic analysis should influence a designer's assignment of certain tolerances, fits, and other parameters.

Let us give a mathematical interpretation of the cost using the regression analysis apparatus.

It is necessary to create a mathematical model that specifies certain mathematical relationships between the dependent quantities yi, i = 1,..., n, which are output characteristics (responses), and the independent variables xj, j = 1,..., m, which serve as input variables or factors. We define a mathematical model as the conditional expectation of a dependent variable for given values of factors:

$$\eta\ (x1,\ x2,\ ...,\ xm) = M\ \{y/x1,\ x2,\ ...,\ xm\}.$$

This equation establishes the change in the average value of the object's response (cost) from the change in factors and is called the regression equation.

Let us give the measured output characteristic or response

$$y(x1, x2, \ldots, xm) = \eta(x1, x2, \ldots, xm) + \varepsilon,$$

where ε are random perturbations with normal distribution, zero expectation, and variance $\sigma2$. It is assumed that the action of a set of random perturbations on an object is equivalent to the action of a single such perturbation. The response $\eta(x1, x2, \ldots, xm)$ is a regression model, which is defined by a specific type of function, expressed both linearly and in any other form, depending on the coefficients included in it. Then, for a random variable yi, we can write

$$yi(x1, x2, \ldots, xm) = \eta + \varepsilon = \sum_{i=1}^{n} \beta_i f_i(x1, x2, \ldots, xm) + \varepsilon$$

where β_i are the weight coefficients of the function $(x1, x2,\ldots, xm)$, which are called regressors.

Limited to a linear class of models, we have

$$yu = \sum_{i=1}^{k} \beta_i f_{iu}, \qquad (4.2)$$

where $u = 1, \ldots, N$ is the observation number; i is the number of factors (Table 4.1).

Table 4.1. Observation table

Time	x1	x2	...	xn	Y
1	x11	x21	...	xn1	y1
2	x12	x22	...	xn2	y2
3	x13	x23	...	xn3	y3
...	...	...	...	...	...
N	x1N	x2N	...	xnN	yN1

Note. The functions f_{iu} correspond to the values X_{iu}.
Let us create a matrix of regressors

$$F = \begin{Vmatrix} f_{11} & f_{12} & \ldots & f_{1m} \\ f_{21} & f_{22} & \ldots & f_{2m} \\ \ldots & \ldots & \ldots & \ldots \\ f_{N1} & f_{N2} & \ldots & f_{Nm} \end{Vmatrix}, \qquad (4.3)$$

in each row of which the values of the function $f_{1u}(x1, \ldots, xm)$ from model (4.2) for a given observation $u = 1, \ldots, N$.

Regressors can be calculated from the original model structure. Since the model (4.2) gives an inaccurate copy of the true character of the function y_u, it can be refined by some function

$$\widehat{y_u} = \sum_{i=1}^{n} b_i f_{iu}, \tag{4.4}$$

called hypothesis y_u and serving as an estimate of the true value of the random variable y_u. To obtain estimates of the coefficient β_i of the regression model, the method of least squares method is used, which allows you to minimize the value $Q = \sum_{u=1}^{N} (y_u - \widehat{y_k})^2$ and get the desired coefficients β_i. The value Q, called the loss (risk) function, can be obtained arbitrarily small.

From the matrix of regressors (4.3) we find the transposed matrix FT. Multiplying the original matrix by the transposed one, we find the information matrix $F^T F$ that allows to obtain the system of normal equations.

When solving the system, we consider the values of the coefficients bi. Thus, we find the function

$$\widehat{y_u} = \sum_{i=1}^{k} b_i f_{iu},$$

that is, we come to the model that $\widehat{y_u}$ is predicted for y_u. The more accurate the estimates are, the more accurately the model $\widehat{y_u}$ defines the original model.

Consequently, the apparatus of mathematical modeling provides a general methodology for constructing various dependencies based on experimental data.

Simulation modeling of the cost and labor intensity of manufacturing parts and units is carried out on the basis of information on factors affecting the cost, including data on materials and own production capacities. By varying these factors it is possible to obtain the optimal value of its cost and labor intensity of production provided that the part meets the specified technical characteristics (Figure 4.3).

Reducing the expenditure of manufacturing parts, and therefore their cost is related with solving optimization problems, usually based on the choice of quality criteria, constraints and optimized (variable) parameters (variable tasks).

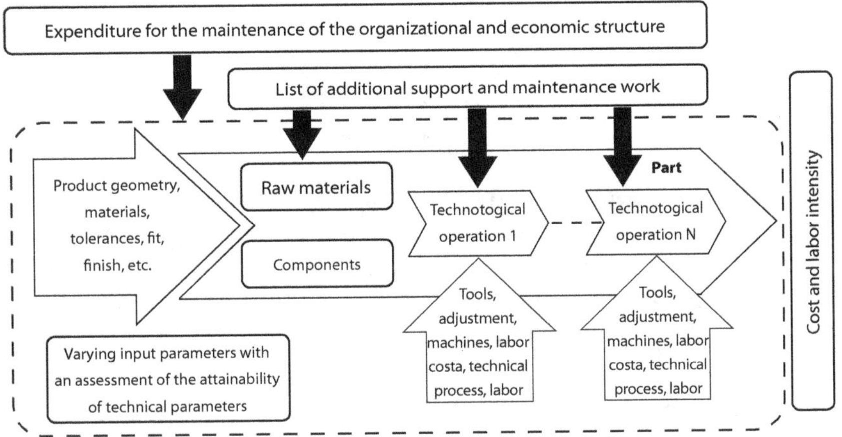

Figure 4.3. Optimization of the cost and labor intensity
of part manufacturing

Let us analyze the method used to solve the problem of cost optimization. By definition, the cost is expressed as the sum of the expenditure of production:

$$C = E_1 + E_2 + \cdots + E_n = \sum_{i=1}^{n} E_i.$$

The cost can be considered as a function of variables E_1 independent of n. In general form, it can be written as

$$y_u = \sum_{i=1}^{n} \gamma_i \varphi_{iu},$$

where φ are arbitrary functions of factors that do not include unknown coefficients, and the experimentally determined function (x1, ..., xn) determines the cost category. The cost function is also the sum of expenditures by the factors that determine it (parameters):

$$\varphi_u = \sum_{i=1}^{m} b_i x_{iu},$$

where u = 1,..., N is the number of observations.

Let us assume that the function φ_u is the expenditure on improving the technical level with set standard, i.e. expenditure is set to improve the quality of this product. In other words, this valuehas a certain relation and dependence on the parameters that make up it.

Similar procedures should also be carried out for other expenditures. Thus, we obtain a system of linear equations:

$$\varphi_1 = b_1 x_{11} + \dots + b_n x_{n1};$$
$$\varphi_2 = b_1 x_{12} + \dots + b_n x_{n2};$$
$$\dots\dots\dots\dots\dots\dots\dots\dots$$
$$\varphi_N = b_1 x_{1n} + \dots + b_n x_{nN}.$$

Here x_{in} are set experimentally, and b_i is unknown.

Solving the system of linear equations with respect to b_i, we obtain the values of the b_i coefficients as correction. Substituting these values into the model describing expenditures, we obtain an equation reflecting the dependence of a variable function on independent parameters.

By imposing conditions and constraints, a functioncan be defined as a target one using one of the optimization methods, and thus its optimum can be found.

Similar operations should be performed with all other φ_u functions in order to find their optimal values.Applying similar reasoning to the model

$$y_u = \sum_{i=1}^{n} \beta_{im}\,\varphi_{iu}; \quad u = 1, \dots, N,$$

we find the β_i values. Substituting these values in the mathematical model that defines y, we identify the optimum of the function y, i.e. the optimally determinedproduction costs at the production stage, assuming that the parameters that affect the cost in the production phase are not correlated with the parameters that affect the cost at the stage of product operation. But the last assumption is not confirmed by practice, since savings at the development stage tend to lead to an increase in the cost of production and operation, and vice versa. However, the dependencies can be very complex.

The considered mathematical model of the cost at the production stage is invariant in nature to any system of determining factors and is independent of the number of observation levels and the number of factors at each level.

Similarly, the optimal cost (C_{opt}) of products at the stages of product development and operation can be obtained.

It's obvious that

$$C_{opt} \leq \sum_{i=1}^{n} E_{iopt}.$$

Already at the stage of forming a project idea, it is necessary to have an idea of the upcoming expendituresconnected to the production and operation of the product.

The concept of life cycle cost (LCC) is one of the most important consumer parameters of high-tech products. In general, the LCC can be described as the costs (expenditures) connected to the development, production and operation of products.

The total LCC is the sum of the costs of product development and production, as well as commissioning and after-sales service (maintaining the product in working condition). When calculating the LCC for new products, it is necessary to consider the expenditures on retraining and advanced training of employees engaged in ctechnological operations with equipment, as well asinevitable losses when the planned profit decreases during the development of new equipment and technology.

Expenditures in the development, production and operation of products C_N are expressed in the following ratio

$$C_N = C_{pp} + C_n N + \overline{C_E} N T_E, \qquad (4.5)$$

where C_{pp} is the cost of product development; C_n is the average specific (per product) cost of a productionbatch; $(\overline{C_E})$ is the average unit cost per T_E time period (per product) for the operation of a batch of products for one year.

It is noteworthy that about 90% of the expendituresinvolved in the production and operation of a high-tech product are determined at the R&D stage. This feature is linked to the high technological complexity of the manufactured product. For this reason, one of the most important principles of the LCC cost concept is forecasting and managing the expenses for manufacturing and operating a new product at the design stage.

It is obvious that the specific (per product) costs $(\overline{C_N})$ of creation and operation during the T3 time periodof the entire batch (including N products) can be expressed by the ratio $(\overline{C_N}) = C_N / N$ and by considering the expression (4.5), the unit costs for the creation and operation during T_3 years of a batch, including N products, will amount to

$$\overline{C_N} = \frac{C_N}{N} = \frac{C_{pp}}{N} + C_n + \overline{C_E}T_E. \tag{4.6}$$

Thus, both total and unit costs for creation and operation depend on the volume of output, determined by the value N.

It should be noted that the unit cost of production of the system also significantly depends on the volume of production.It has to do with the fact that an increase in the number of manufactured products results in decrease of the average cost of its unit, due to the distribution of production preparation costs, costs for new equipment and production facilities, personnel training, etc. Moreover, experience is accumulated in the production process, technological processes are improved, which increases labor productivity and reduces expenditure per unit of production. The relationship connecting the cost of the j-th item in a batch of volume N with the cost of the first item C_1 can be considered

$$C_j = C_1 e^{-\theta_1(j-1)}, \tag{4.7}$$

where θ_1 is the constant coefficient.

Using the expression (4.7), the average cost of the product in such a batch can be calculated

$$C_{av} = \frac{1}{N}\sum_{j=1}^{N} C_j = \frac{C_1}{N}\sum_{j=1}^{N} e^{-\theta_1(j-1)} \approx \frac{C_1}{N}\int_1^N e^{-\theta_1(j-1)}\, d_j =$$

$$= \frac{C_1}{N}e^{\theta_1}\frac{e^{-\theta_1 j}}{-\theta_1}\Big|_1^N = \frac{C_1}{N\theta_1}e^{\theta_1}(e^{-\theta_1} - e^{-\theta_1 N}),$$

from which $C_{av} \approx \dfrac{C_1}{\theta_1 N}[1 - e^{-\theta_1(N-1)}].$

Now, to determine the coefficient θ_1, the sum of the optimal cost values calculated according to the proposed method at all stages of the product life cycle can be used.

The modern practice of managing the cost of new technology products at all stages of the life cycle involves the active use of innovative solutions in the form of software products aimed at solving the problem of designing products for a given cost and competitiveness in the market, which change the existing paradigm of the product life cycle. In this regard, new economic mechanisms are needed to improve the efficiency of planning and using all types of resources of an organization in the process of creating unique products with high competitive advantages and optimal cost achieved as a result of rational economic support for the life cycle of its development and production.

4.3. THE ROLE OF DIGITAL MODELS AND DIGITAL MODELING IN COST OPTIMIZATION AT THE STAGES OF THE PRODUCT LIFE CYCLE

Optimization of costs at the stages of the life cycle of promising products is a complex economic and technical problem, since when designing and creating promising products, a large number of random factors, the forecast of which causes difficulties connected with innovations must be taken into account. Moreover, traditional methods of cost optimization are not applicable due to the large stochasticity of the optimization problem. Stochastic optimization methodstend to be effective only in the case of mass application of optimal solutions, since these methods optimize only on average, andit is necessary to reduce costs in individual when creating promising products.

The use of digital models or digital mockups of products plays a key role to optimize costs at the stages of the life cycle of promising products. A modern digital model of a product consists of a large set of various electronic documents describing a promising product. In particular, the digital model contains electronic document flow, and most importantly – a control system for product components and three-dimensional models of product components. Also, the digital model of a product includes various technological and production data.

We will show how the use of a digital model allows solving the problem of cost optimization at the stages of the life cycle of a promising product. To do this, we will describe an economic and mathematical model of cost optimization in the life cycle of promising products.

Let the life cycle of a promising product consist of N different stages. We will denote each of these stages as

$$K_1, K_2, \ldots, K_N.$$

We can consider the following stages:

- marketing research;
- scientific and search works;
- planning of manufacturing proccesses;
- purchase of materials and components;
- creation of the main components;
- product assembly;
- installation and commissioning;
- testing;
- acceptance;
- intended use;
- maintenance and repair;
- recycling.

We will consider the case when the stages of the life cycle are performed sequentially, without parallelization. The implementation of each stage requires certain costs. This scheme is shown in Figure 4.4.

In figure 4.4 K_n denotes the stages of the life cycle of a promising product, and F_n denotes the necessary funding (expenditure) for the implementation of K_n stages, where n = 1,..., N.

In figure 4.4, the stages of the life cycle of a promising product are indicated by K_n, and the necessary funding (expenditure) for the implementation of the K_n stages are indicated by F_n, where n = 1,..., N.

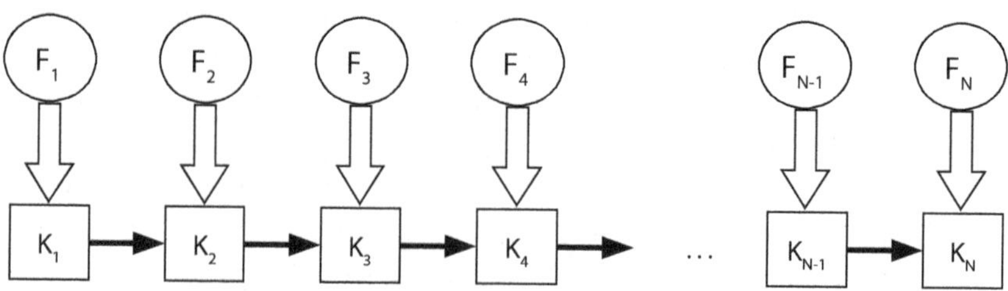

Figure 4.4. Funding scheme for product life cycle stages

In what follows, K_n will denote not only the stage of the life cycle itself, but also the result obtained after completing this stage.

The total expenditure of implementing the product life cycle is determined by the formula

$$F = (F_1 + \Phi_1) + (F_2 + \Phi_2) + \ldots + (F_{N-1} + \Phi_{N-1}) + F_N,$$

where the values F_n represent the additional expenditures that arise when preparing the product for stage $(n + 1)$, after the implementation of stage n. In particular, these expenditures arise because the parameters obtained during each stage may differ from the planned values. For the case under consideration, these expenditures are random. Thus, we have a stochastic programming problem.

To formulate the problem statement for stochastic optimization, we introduce a probability space

$$\Pi = \langle \Omega, B, P \rangle,$$

where Ω is the space of elementary outcomes; B is the sigma algebra of events on the set Ω; P is the probability measure of events from the set B. For each elementary outcome $\omega \in \Omega$ we have the following optimization problem

$$F(\omega, u_1, u_2, \ldots, u_N) = \Sigma \, (F_n(u_n) + \Phi_n(\xi_n(\omega), u_n)) \to \min. \qquad (4.8)$$

Here, through $u_1, u_2, \ldots, u_N$, controls are indicated at each stage of the life cycle of a promising product. They are selected from a variety of possible controls

$$u_n \in U_n,$$

where U_n is the set of possible or valid product lifecycle controls. These sets must not be empty. By $\xi_n(\omega)$ we denote random variables.

In practical problems, the conditions are always satisfied that these random variables $\xi_n(\omega)$ have finite first and second moments:

$$|E[\xi_n(\omega)]| = |E_n| < \infty,$$

$$D[\xi_n(\omega)] = D_n < \infty.$$

Here, $E[\xi_n(\omega)]$ denotes the mathematical expectation of the random variable, and $D[\xi_n(\omega)]$ denotes the variance of the random variable.

The direct solution of the optimization problem (4.8) is practically impossible, since it contains random variables. Standard methods for solving stochastic problems involve replacing problem (4.8) with a problem in which random variables are replaced by their mathematical expectations, which characterize the average values of these random variables. At the same time, mathematical expectations, like variances, are nonrandom quantities.

Thus, the stochastic programming problem (4.8) can be replaced by a mathematical programming problem using the following formula:

$$F(u_1, u_2, \ldots, u_N) = \Sigma\ (F_n(u_n) + \Phi_n(E_n, u_n)) \to \min. \qquad (4.9)$$

Problem (4.9) can already be solved by standard methods of optimization of nonlinear functions.

However, for the considered case of cost optimization at the stages of the life cycle of a promising product, this approach will lead to large errors, since replacing a random variable with its mathematical expectation is justified only if this optimization problem is solved many times.

The value of D_n is an estimate of the accuracy for solving a stochastic optimization problem. The smaller the variance value, the smaller the spread of the random variable values relative to the mathematical expectation.

We will show the role that the use of a digital product model plays in solving the problem of cost optimization at the stages of the life cycle of promising products. The digital model allows getting an accurate picture of the product itself at all stages of its design, as well as the components of the product. Therefore, when solving problems of cost optimization, it is necessary to have reliable information about the intended product. This information can be used to reduce the variance of random variables that affect expenditures.

We introduce the concept of a digital benchmark model of a promising product at the stages of the life cycle. A digital benchmark is a set of information that describes the expected characteristics of a product at each stage of the life cycle. The mathematical expression for a digital benchmark is as follows:

$$X = (X_1, X_2, \ldots, X_N),$$

where each element X_n is a k-dimensional vector in the parameter space R^k. We will assume that we have functions that can assess the compliance of the results of the product life cycle stage with the planned parameters that are proportional to the digital benchmark. These functionals will be denoted as follows:

$$\Psi_n = \Psi_n(K_n, X_n), n = 1, 2, \ldots, N.$$

We assume that $\Psi_n \geq 0$ have the meaning of a measure of the discrepancy between the result K_n and the parameters X_n. In the case of a complete match, we have $\Psi_n = 0$.

Digital benchmarks obtained from the digital model of the product, allows to adjust the performance of each stage of the life cycle to solve an extreme problem

$$\Psi_n(K_n(v_n), X_n) \to min, n = 1, 2, \ldots, N. \tag{4.10}$$

Here, v_n denotes the control actions at the n-th stage of the life cycle of promising products, which are selected in such a way as to provide a solution to the problem (4.10) and significantly reduce the values of the functions $F_n(\xi_n(\omega), u_n)$, which will optimize costs at all stages of the product life cycle. In addition, the use of digital benchmark in the life cycle makes it possible to seriously reduce the variance of random variables $D[\xi_n(\omega)]$. This serves as a mathematical basis for solving the stochastic problem (4.9), when the random variables included in the problem (4.8) are replaced by their mathematical expectations.

Thus, the use of two-stage cost optimization at the stages of the life cycle of promising products allows us to formulate the following effective methodology for solving the problem of cost optimization in the implementation of the product life cycle.

1. Introduction

When implementing the life cycle of promising products in knowledge-based organizations, it is necessary to solve the problem of optimizing financial and other resources. The specificity of promising products is that when performing this task, a large number of different uncertainty factors arise as a result of the inevitable risks connected with the introduction and use of innovative technologies.

The digital model of a product that corresponds to the conditions of the modern digital economy is the most important tool for optimizing costs at the stages of the life cycle. Having a digital model allows using various methods to effectively optimize costs at stages of the life cycle.

This technique consists in the fact that digital benchmarks of the life cycle of a promising product are used as a digital tool. These benchmarks provide a digital description of the planned product characteristics at each stage of the life cycle, as a result of which it is possible to form optimal solutions for cost savings.

2. General provisions

The method of using the digital model is designed to solve the most important problem arising when implementing the stages of the life cycle of high-tech products, such as achievement of the specified performance characteristics of the product, which should ensure the high competitiveness of the product. In doing so, innovative technologies are used, which actually allow getting competitive advantages. However, in this situation, there is another important problem, which is to optimize (reduce) costs at all stages of the life cycle of a promising product. To solve this problem, there are various traditional methods. However, when creating promising products, the most important tool is the digital model or digital mockup of the product.

The use of a digital model of a product allows to receive up-to-date information about a product and its components at all stages of the life cycle; this information, in turn, is used for cost optimization.

3. Goals

The goal of the method of using the digital product model in solving the problem of cost optimization at the stages of the life cycle of promising products is to significantly reduce the amount of funding using modern digital technologies, as well as to form optimal solutions when implementing the stages of the life cycle of promising products.

4. Tasks

This method of using a digital model solves the following tasks:

- shaping management decisions in the implementation of the life cycle stages of promising products;
- reduction of financial expenditure in the implementation of product life cycle stages;
- optimal management decisions that allow maintaining the given characteristics of the product at all stages of the product life cycle with minimal costs.

5. Indicators determined using the methodology

The method of using a digital product model in solving the problem of cost optimization allows to determine the following indicators:

- management decisions at each stage of the product life cycle to maintain the specified product characteristics;
- product lifecycle management solutions that ensure cost optimization;
- optimized costs at the stages of the life cycle of promising products.

6. Initial data

The following data are used as the initial data of the method of using the digital product model in solving the cost optimization problem:

- digital model;
- description of the life cycles of a promising product;
- basic funding for life cycle stages;
- additional costs when preparing the product for the next stage;
- random factors affecting expenditure on the stages of the life cycle;
- many possible management decisions

7. Methodology and algorithms

The method of using a digital product model in solving the problem of cost optimization at the stages of the life cycle of promising products is implemented in several stages.

Stage 1.

Building digital benchmarks of the product life cycle. Using the digital model of the product, digital benchmarks are built for each stage of the life cycle. A digital benchmark refers to the expected performance of a product after the implementation of each stage of the product life cycle.

As a result of this stage, a sequence of vectors is formed

$$X = (X_1, X_2, \ldots, X_N),$$

where each element X_n is a k-dimensional vector in the parameter space R^k, called the digital benchmark of a product..

Stage 2.

Construction of numerical functionals for assessing the compliance of the actual results of the implementation of the stages of the product life cycle with numerical characteristics according to digital benchmarks. These functionals are denoted as follows:

$$\Psi_n = \Psi_n(K_n, X_n), n = 1, 2, \ldots, N.$$

Stage 3.

The solution of the first optimization problem, which consists in choosing the optimal control action at each stage of the life cycle in order to minimize the functionals Ψ_n.

For this it is necessary to solve the extremal problem (4.10). As a result we obtain optimal (or close to optimal) management decisions

$$v^*_1, v^*_2, \ldots, v^*_N.$$

Stage 4.

Identify random factors that affect expenditures at the life stage. It is necessary to construct random variables based on the analysis of random factors that affect the increase in expenditures at the stages of the life cycle of promising products.

$$\xi_1(\omega), \xi_2(\omega), \ldots, \xi_N(\omega).$$

For each random variable, its distribution functions should be found.

Stage 5.

For each random variable, calculate the mathematical expectation and variance:

$$E_n = E[\xi_n(\omega)],$$

$$D_n = D[\xi_n(\omega)], n = 1, 2, \ldots, N.$$

It is necessary to make sure that the obtained values of the variance D_n are sufficiently small. Otherwise, it is necessary to return to stage 3 to solve the first optimization problem.

Stage 6.

Solution of the second optimization problem (4.9), which consists in finding optimal management solutions for the following extremal problem

$$F(u_1, u_2, \ldots, u_N) = \Sigma\ (F_n(u_n) + \Phi_n(E_n, u_n)) \to \min.$$

After solving this problem, find the optimal values of management decisions

$$u^*_1, u^*_2, \ldots, u^*_N,$$

and also determine the value of the optimized life cycle costs of a promising product using the following formula:

$$\Sigma\ (F_n(u^*_n) + \Phi_n(E_n, u^*_n)).$$

As a result of the indicated stages of the methodology for using the digital model of the product in solving the problem of cost optimization at the stages of the life cycle of promising products, we obtain an optimal plan for implementing the life cycle of promising products.

From the above, it appears that there is a need to create and develop control systems for a cost-effective process of launching new products into production, for which an economic and mathematical model of cost optimization at the stages of the life cycle of promising products is built based on the use of a digital product model, which allows assessing management decisions, reducing overall costs; the concept of a digital benchmark based on a digital product model is also introduced, through which it is possible to optimize costs due to a more accurate correspondence of the results of the implementation of life cycle stages and the planned characteristics of the product, which are directly related to the economic development of knowledge-based organizations.

CHAPTER 5. ECONOMIC MECHANISM FOR MANAGING THE COMPETITIVENESS OF PROMISING HIGH-TECH PRODUCTS

5.1. APPLICATION OF DIGITAL MODELS FOR OPTIMIZATION OF PRODUCTION AND TECHNOLOGICAL PROCESSES

Modern organizations are increasingly using the latest methods of design, production and release of products related to the widespread use of digital technologies. Digital information technologies, virtual reality technologies and the creation of digital twins, technologies based on artificial intelligence systems, 3D technologies and many others that allow to automate the processes of creating and manufacturing products, reduce labor intensity and cost, and increase labor productivity are most actively developed. Using digital models of mechanisms, processing units and personnel, it is possible to simulate the work of entire production sites and workshops, as well as linked logistics chains, i.e. to determine in advance the most optimal design and technological solutions, labor intensity of manufacturing a unit of production, various production risks. The optimal parameters of the production and technological process must be determined to ensure the proper quality of products, an acceptable level of labor intensity and cost of the work performed.

Automation of procedures for optimizing production and technological processes is one of the main tasks which is solved with the use of digital technologies of development and production based on the creation of digital twins of a product. Using "smart" intelligent methods by means of simulation mathematical modeling, it becomes possible to automatically determine, for example, the parameters of the technological process. In the paper[19], a digital twin of 3D printing on CNC machines was proposed, based on

---

19 Kabaldin Yu.G., Kolchin P.V., Shatagin D.A., Anosov M.S., Chursin A.A. Digital Twin for 3D Printing on CNC Machines // Vestnik mashinostroeniya. – 2019. – № 7. – Pp. 47–49.

which depending on the material of the part, the specified values of tolerances for the element size, yield stress and other parameters of the workpiece, as well as the parameters of technological equipment (width of a single bead, height of a single roller, etc.), the parameters necessary for 3D printing of a part and its further processing by a CNC machine (printing speed, type of gas mixture, etc.) are determined, while maintaining the quality characteristics of the part specified by its developers.

The complexity of the task of optimizing production and technological processes is linked to a large number of parameters, on the basis of which design and technological decisions are made: material choice, setting tolerances and fits, detection of equipment and processing methods, etc. By varying these factors, provided that the part meets the specified technical characteristics, it is possible to determine the optimal manufacturing method. Modern production provides a fairly large range of options for manufacturing products and their components, which is constantly expanding due to the growth of technological capabilities, achievements of science and technology. The modern way to choose the optimal option for manufacturing a part is to use intelligent data analysis methods, which will be proposed in this paper.

In the conditions of digitalization of the processes of design, preparation and product manufacturing for managing the organization of the technological process, it becomes possible to use modern methods of intelligent information processing. However, most of this information is not available for direct analysis by decision-makers. One of the reasons is the high dimension of information flows, which is manifested in the fact that any information events are simultaneously described by many different indicators. Moreover, these indicators are constantly changing over time. For a true assessment of the state of the system, it is necessary to simultaneously analyze not individual indicators, but the entire set of multivariate data. This raises am urgent task of the construction of effective methods and information systems for the operational analysis of multidimensional flows of production and technological data for managing the organization of production and technological work.

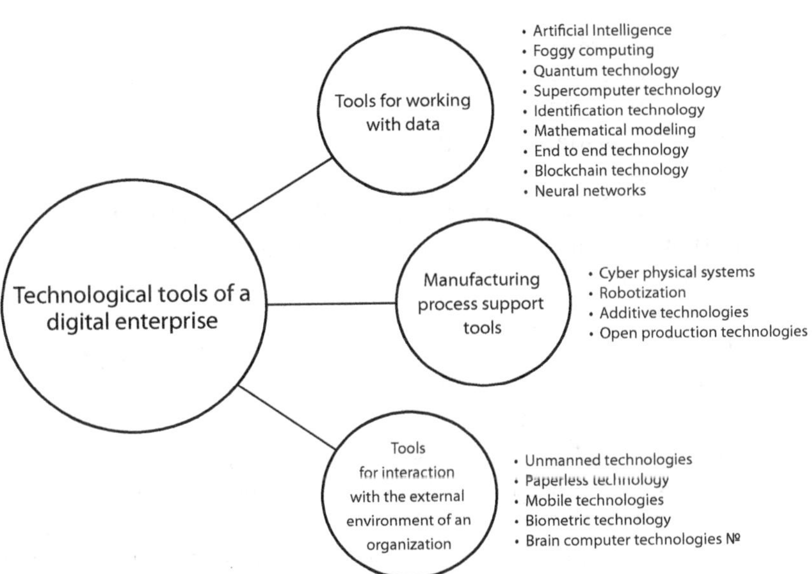

Figure 5.1. Technological tools of a digital enterprise

Modern methods for solving this problem are connected with the digitalization of the main processes for product development. Big production data and analytics are becoming one of the most important assets and are being used vigorously in managing the organization of work on the development of new products. Analysis of the accumulated data allows automating and improving operations within the functioning of a single digital enterprise, whose work is organized through the use of complex technological tools, represented by a set of advanced technologies used in various areas of economic activity. We will highlight three main complex tools that are already actively used in practice. These include: tools for working with data that is becoming one of the most important assets of an organization; tools for ensuring the production process; tools for interacting with the external environment of an organization. The complex of these technological tools is shown in Figure 5.1.

Based on Figure 5.1, it can be seen that the tools for working with data are represented by such technologies as artificial intelligence, fog computing, end-to-end, quantum and supercomputer technologies, identification technologies, blockchain, neural networks, etc. The combination of these technologies allows for effective management and use of big data about the internal and external environment of the organization, on which the process of managing design and technological work can be built. This toolkit makes it possible to use past experience to shape future decisions using advanced self-learning algorithms. In addition, decision-making models adapt to new conditions when new data is received. The achievement of high accuracy of their processing and minimization of errors that appear due to the human factor is an important advantage of using these technologies in the process of data management.

Let us consider the issue of managing the organization of a technological process through the use of modern methods of intelligent data processing based on neural networks and building a digital model (twin) of a technological process. In this case, the term «digital model» refers to a mathematical model of adaptive control of the parameters of the technological process depending on the parameters of the part (workpiece) and the technological equipment used.

Fig. 5.2 presents a neural network model (twin) of 3D printing on CNC machines [20]. Through a neural network model, procedures were carried out for setting the recommended 3D printing parameters and maintaining its modes at a level sufficient to obtain the structural and mechanical characteristics of the final product specified by the developer.

In practice, neural networks can work with dozens of input and output parameters, providing the choice of the optimal output set based on processing the set of initial data. Neural networks are widely used in problems of managing multi-parameter processes, when effective management decisions are developed as a result of analyzing heterogeneous data. Big data generated by economic processes (including production processes) is the result of the widespread digitalization of the modern economy, accompanied

---

20 Kabaldin Yu.G., Kolchin P.V., Shatagin D.A., Anosov M.S., Chursin A.A. Digital Twin for 3D Printing on CNC Machines // Vestnik mashinostroeniya. – 2019. – № 7. – Pp.. 47–49.

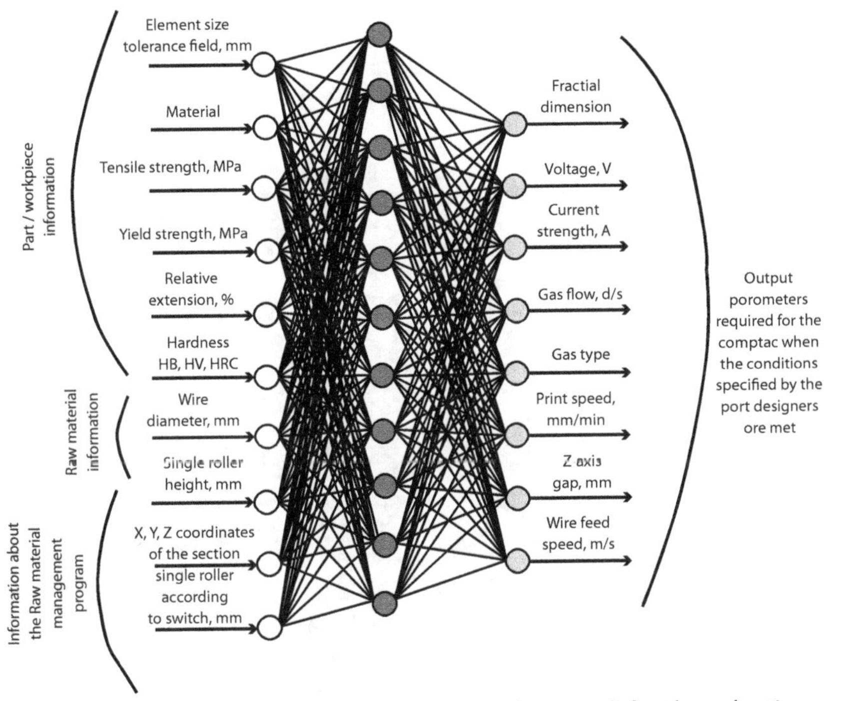

Figure: 5.2. Application of a multilayer neural network for the selection
of process parameters

by a significant information flow in any production or organizational activities of organizations.

A neural network model called Kohonen self-organizing maps is ideal for solving the problem of simultaneous deep analysis of multidimensional information flows. This method allows building visual graphic maps that reflect the distribution of multidimensional data according to their "internal structure", building visual marked maps of various states of the system.

The generated marked maps can be used to classify and present multidimensional information, and machine learning methods based on Kohonen self-organizing maps automatically analyze the original data. After building self-organizing maps, they are also used as a visual indicator of the current state of a complex system. To do this, at each moment of time, a point corresponding to the current

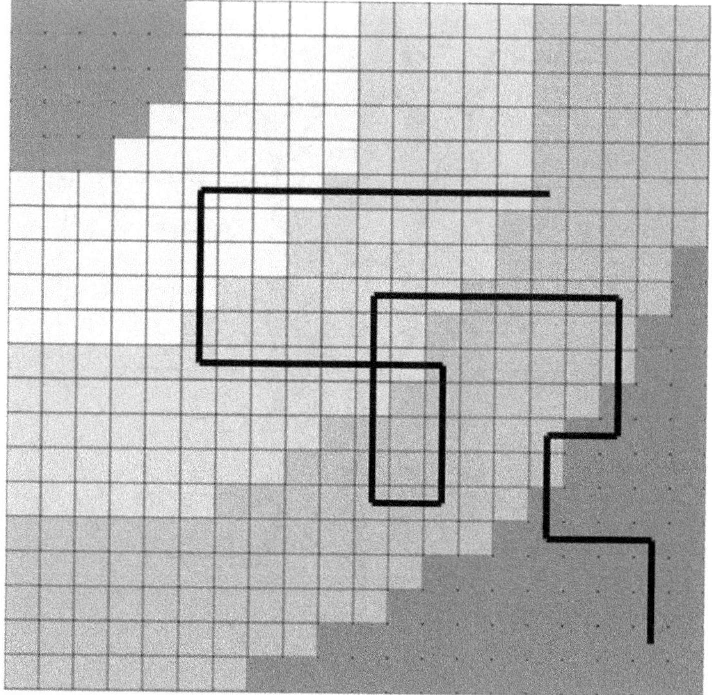

Figure 5.3. A Kohonen map with the trajectory of system states

state of the system must be plotted on the map, which will make it possible to construct a trajectory of movement of the state of the system on the Kohonen map. Analyzing the behavior of this trajectory, it is possible to identify dangerous trends in the development of the system, take proactive management decisions. Kohonen map allows not only to analyze the current state of affairs, but also to evaluate the management decisions made in an operational mode in real-time.

Figure 5.3 shows an example of a marked Kohonen map, where the area corresponding to the state of the system is highlighted in dark gray in the lower right corner of the figure, and the dangerous states of the system are highlighted in black (upper left corner). The black line is the trajectory of the system states on the Kohonen map. Some of the trajectory changes are caused by control decisions aimed at correcting the dangerous situation.

The method of analysis of multidimensional information flows based on the construction of Kohonen self-organizing maps can be used to study the behavior of the following complex systems in various fields of activity;

- production and technological audit of an organization as a large techno-economic system. Its functioning is connected with a large number of techno-economic indicators that describe the state of an organization. Moreover, only a comprehensive analysis of all data allows assessing its true state, as well as to detect dangerous economic trends in its activities;
- a complex technical system, which is also described by a large number of indicators, which together help to detect possible failures in the functioning of the system;
- a multi-magnetic organizational and technical system. Many technical systems, such as automated production lines, are described by large data streams, which can include both production and technological, as well as organizational and managerial information. The efficiency of the entire system is not limited to the successful operation of only each component, it requires the correct coordination of the entire system.

Thus, data mining is a powerful tool that can be effectively used at all stages of creating highly competitive products and become an important factor in the development of its optimal cost and labor intensity of production (Figure 5.4).

The chain of creating a promising product in an industrial enterprise is divided into the following phases, each of which can be characterized by large data sets (Figure 5.5):

- product development;
- production planning;
- pre-production and engineering;
- actual production;
- maintenance and repair.

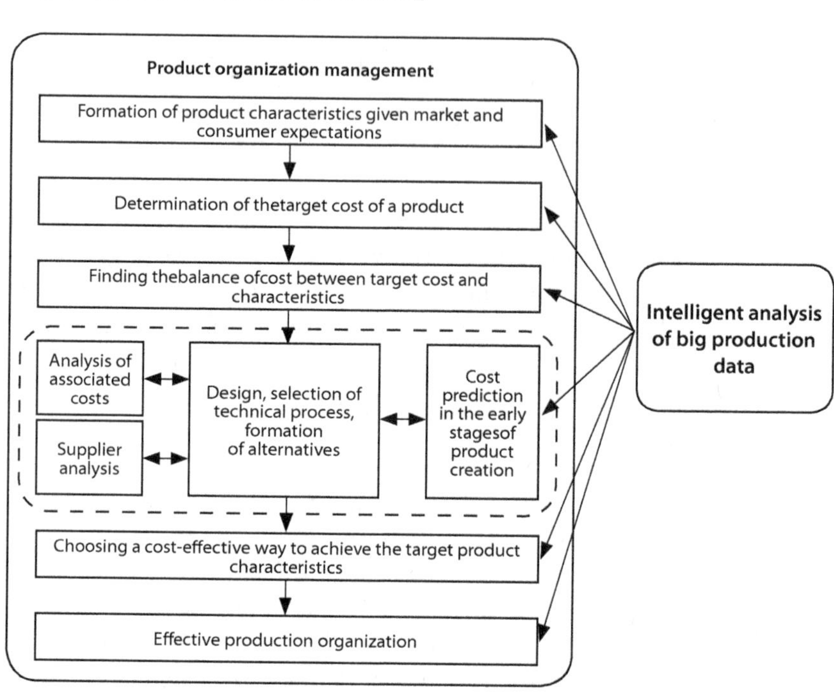

Figure: 5.4. Application of data mining at all stages of product management

Figure: 5.5. Product development chain

At each of these phases, depending on the specific production, different sets of components, tools, possessing units are used and specific tasks are solved by constructing a digital twin of each phase. At all stages, there is interaction with internal and external

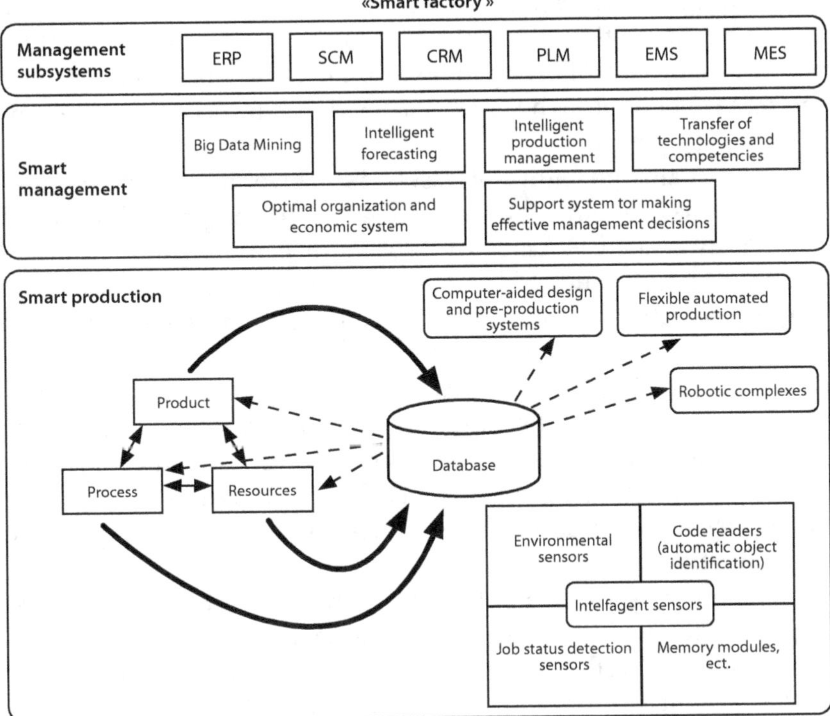

Figure: 5.6. Digital business processes of a "smart factory"

factors (e.g., suppliers), which is a separate optimization task based on processing big data about suppliers. As part of the digitalization of business processes, a "smart factory" is created as a reflection of this chain in the virtual world, consisting of digital twins of objects and processes in the real world (Figure 5.6). Then, using information methods of working with digital twins, the forecasted techno-economic effect is calculated, which can be obtained as a result of building products according to the developed and digitally modeled image of the product and the stages of its creation.

Let us review the last task in more detail. Based on a certain number of indicators characterizing the technical level of production, product quality and labor productivity, it is possible to construct mathematical models, including both characteristics of the state of production (progressiveness of the technological environment, the degree of automation of production, capital productivity, capital-labor ratio, etc.), and allowing to optimize the impact of innovation on the achievement of high values of specific techno-economic parameters of products.

The main indicators of assessment of efficiency of production serve as economic indicators (e.g., profit and profitability) and the closely related technical indicators of production, controlling quality, technical and technological conditions of production and output indicators of innovation. A significant contribution to the efficiency of the choice of innovative technologies is made by data on manufacturing technologies (e.g., taking into account the preliminary stress-strain state and thinning, deformation, etc. of parts after technological processes). Accordingly, the assessment of innovation of technologies can be carried out taking into account the characteristics of a particular production based on a digital twin of production.

Combining the digital twin of a product and the digital twin of production within a single digital model leads to a "smart" digital twin, which is a reflection of several stages of the product life cycle (development, pre-production and production).

Based on such a twin double, the issue of effective organization and technical equipment of production can be solved, taking into account modern innovative technologies to ensure the production of competitive and demanded products on the market at minimal costs. The mathematical model underlying such a digital twin is:

$$\sum_{i=I}^{I} C_i(t) = \sum_{i=I}^{I} M_i(t) + S_i(t) + D_i(t) + O_i + C_{dev} \to min$$

where $C_i(t)$ is production costs of the i-th product;

$M_i(t)$ is expenditure for materials and components of the i-th product;

$S_i(t)$ is salary for the i-th product;

$D_i(t)$ is depreciation of fixed assets for i-th production;

O_i is operating and repair costs of funds for the i-th product;

C_{dev} is costs for the new innovative technology learning.

The expenditure for raw materials, materials and components, taking into account time and reducing consumption rates, are defined by statistical ratios obtained for similar products

$$\sum\nolimits_{i=I}^{I} M_i(t) = mT(t),$$

where T is the specific weight of materials and components in the planned volume of commercial output T (t).

Fig. 5.7 shows a diagram of a digital twin of the organizational and economic system of production.

Production costs for the manufactured and mastered part of the product are for the i-th product

$$C_i(t) - M_i(t) = S_i(t) + A_i(t) + O_i + C_{dev} = \Phi_{oni}(t) +$$

$$+ \Phi n_{dep} + \Phi\xi + F\,a\Delta = S_i(1 + \varphi n_{dep} + \varphi\xi + \varphi a\delta), \qquad (5.1)$$

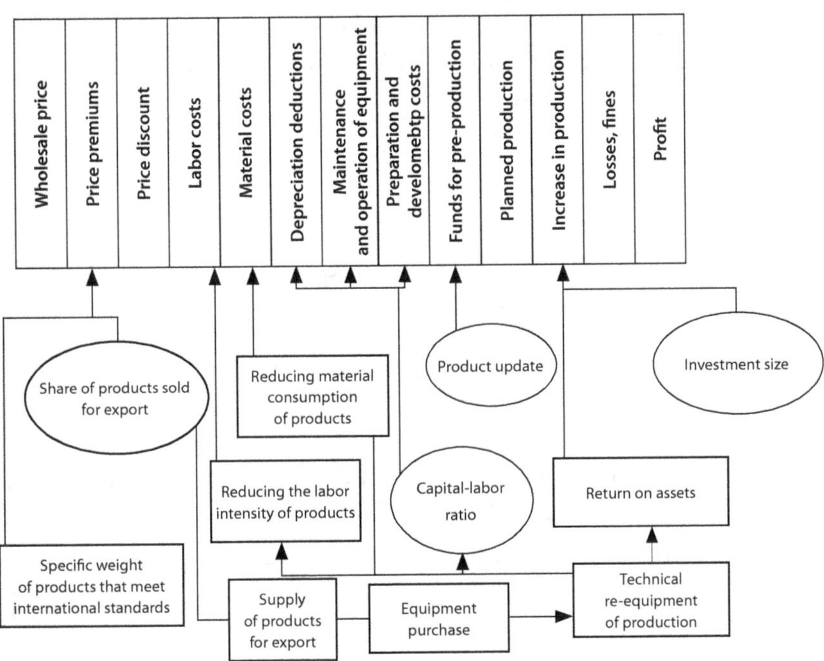

Figure 5.7. Diagram of a digital twin of the organizational and economic system of production

where Φ is cost of fixed assets;

π_{dep} is depreciation rate;

ξ is share of costs of operating funds;

a is share of the active part of fixed assets;

Δ is growth rate of new products;

$\varphi = C/L$ – capital-labor ratio.

Given that

$$C = Li\, t_{rate},$$

where Li is the labor intensity of products; t_{rate} is the average tariff rate, and the labor intensity of products is the inverse of labor productivity, which is associated with the capital-labor ratio:

$$Li = A/P = A/(\varphi\psi) = A/(V\varphi^{v}) = B\varphi^{-v},$$

whereA, V, v, B are statistical coefficients; P, φ are labor productivity, capital productivity.

Statistical coefficients can be determined by known methods. We get an expression for determining the production costs for the manufactured and developed part of the product

$$C = B\varphi^{-\gamma}\tau_{rate}\left[\frac{1}{\varphi(n_{rate}+\varepsilon+\propto\Delta)}\right], \qquad (5.2)$$

which should strive for the minimum to obtain the maximum profit.

Extremum condition

$$\frac{\vartheta C_1}{\vartheta f_{\scriptscriptstyle 6}} = 0$$

φ gives the optimal value of capital-labor ratio $\varphi$and estimated indicators of the technical level of production

$$Bt_{rate}(-V)\varphi^{-V-1}+(1-V)Bt_{rate}\,\varphi^{-V}(n_{dep}+\xi+a\Delta)=0, \qquad (5.3)$$

$$\varphi[-V\varphi+(1-V)(n_{dep}+\xi+a\Delta)]=0, \qquad (5.4)$$

$$\varphi=\frac{V}{(1-V)(n_{rate}+\xi+a\Delta)}. \qquad (5.5)$$

The maximum value of profit is determined by the optimal indicator – the capital-labor ratio φ and is calculated through the indicator of the elasticity of labor productivity from the capital-labor ratio v, standards of depreciation charges n_{dep} and costs for operation and equipment repair ξ, specific weight of active part a and update Δ.

Optimal capital-labor ratio is related to optimal capital productivity ($P^* = \varphi\psi^*$), which can be expressed as

$$\psi = \frac{mP_{eq}K_{sh}K_{i-sh}a}{C_1},\qquad(5.6)$$

where P_{eq} is equipment performance;

K_{sh} is number of machine shifts of equipment operation per day;

C_1 is cost of a piece of equipment;

K_{i-sh} is coefficient of intra-shift fund of equipment operation time (loading);

$K_{sh}K_{i-sh} = K_{ut.}$ – equipment utilization rate – characterizes the organizational level of production, complementing the indicators of the technical level v, n_{dep}, ξ, a, Δ.

Equipment performance depends on the degree of its automation K_{aut} and power availability Po_{av}.

The proposed digital twin of the organizational and economic system of production helps to reveal the relationship of production efficiency indicators (profitability, labor productivity, etc.) with indicators of the organizational and technical level of production and products (material consumption of products, capital-labor ratio, mechanical-labor ratio, level of equipment automation and its power-to-weight ratio, equipment utilization rate etc.) and make a selection of innovative technologies, the development of which corresponds to the economic capabilities of the organization and allows to provide the necessary technical characteristics of the product.

The modern practice of designing and organizing production and technological processes involves the active use of digital economy technologies, which are effective tools for big data analysis of production information, as well as digital models (digital twins) developed on their basis. Such a production management system is the basis for creating a "smart factory", i.e. a digital production management system that has the following characteristics:

- building digital intelligent platforms, unique ecosystems of advanced digital technologies. Based on predictive analytics of big data, the platform approach allows to unite geographically distributed participants in the design and production of a product of the future (create so-called "virtual enterprises" in order to combine the best technologies and competencies to create a promising product at a given cost with high quality characteristics), increase the level flexibility and customization given customer requirements;
- digitalization of the entire product life cycle (idea, design, production, operation, service, and disposal). The later changes are made, the greater their cost, and therefore the center of gravity shifts towards design processes, in which the characteristics of global competitiveness or high consumer requirements are laid, given the competitive price;
- development of the system of digital models of both new designed products and production processes on the basis of the proposed digital twins. Digital models must have a high level of adequacy to real objects and real processes (convergence of the material and digital worlds that generate synergistic effects).

The proposed digital models of products and production and technological processes are aimed at optimal management of the processes of product creation and production using modern intelligent cyber-physical and cyber-economic systems that create an integrated digital platform for supporting effective solutions for product lifecycle management with appropriate information support.

5.2. ASSESSMENT OF THE COMPETITIVENESS OF HIGH-TECH ORGANIZATIONS

The key methodology for analyzing the competitiveness of an organization is the "Porter's five forces analysis", i.e. the five forces of competitive advantage, such as the analysis of the threat of new players, the market power of consumers, the market power of suppliers, the threat of substitute products and the level of competition. However, today many economists-researchers believe that this model is outdated and requires expansion and refinement. It is proposed to modify Porter's model for a much clearer, more precise and complete analysis of the forces that determine the role of competition in modern high-tech industry, taking into account the following four new factors that are paramount to guarantee competitiveness.

1. *Computerization.* It is considered to be the most important in the context of the current vision of business and manifests itself in the threat of lagging behind the increasingly progressive computerization of the market environment. This means that organizations compete both in their own industry and in several industries at the same time. A number of scientists consider four properties of computerization, thanks to which it is possible to assess the competitive advantages of an organization: infrastructure, digital input, digital processing and digital output. Let us consider them in detail:

- the quality and level of technology used in computer networks and when working with the Internet are the most important defining properties of the infrastructure;
- when placing orders, the use of computerized information processing and digital processes for automated order processing (electronic transactions and communications, computer networks) determines digital input;
- a particularly significant element that determines the level of integration of external and internal processes is digital processing. For example, it includes technologies for planning and distributing funds of an organization, CRM (customer base interaction management model), management of production and service itself, expanding knowledge and information in the organization to ensure that certain divisions and functions of the organization are connected. In addition, external integration is electronic data transmission, supply chain management, etc.;
- the final element to prioritize digital processing in sales as well as automation of sales processing is digital output.

A stronger mutual integration of these four elements leads to the best position of an organization in the application of computerization to obtain long-term competitive advantage. According to some experts, it is the significant market players who most often achieve complete excellence in computerization. The higher the degree of computerization proficiency in an industry, the fiercer the competition and the more serious efforts will be required to achieve the desired degree of competitiveness in these four areas.

2. *Globalization.* If just a few decades ago, in order to have a competitive advantage, it was enough to meet the requirements of price and quality, today, in order to be a competitive leader, it is necessary to maintain a global, widely available worldwide network of partners, allowing the customer to purchase a full range of services anywhere in the world requiring the development of deep and long-term relationships with the customer. To use this indicator in practice, a method of measuring this indicator is necessary; however, there is no well-known method yet. Some authors suggest that more globalized countries have lower tax cost and public expenditure. It is obvious that organizations seeking to seize the "power" of globalization should take into account the fact that for competitive superiority, business can be started in a developing country with less fierce competition among international organizations, compared to a developed one.

3. *Deregulation.* According to the American economist A. Downs, current trends are a sharp reduction in state regulation, for example, in the field of telecommunications, air transport, and banking. Despite the fact that there are no clear criteria for measuring deregulation, it is most dependent on local governments and their decisions, changing significantly in different countries. It is also impossible to predict real trends that could mitigate or intensify government intervention and influence. These processes depend on the political and economic state of a country and region. For an organization, there are two ways of action to consider: adjust to local legislation, or choose a country with a more appropriate one. It should be noted that deregulation is not always optimal for business. For example, in Chile, deregulation has created an extremely hectic and volatile market, especially in the telecommunications sector.

4. *Competition and technology.* Global competition in the modern world has become a new reality. Technological progress has forced the confusion of external circumstances that organizations

have to face. The growth rate of technological innovations begins to increase, which leads to the replacement of the once dominant technologies by new technologies, technological primacy is shifting between states, and industry is changing, and in some cases even disappears. High-tech industry products are designed and manufactured with increasing involvement of digitalization, robotics and new materials. For example, e-commerce has changed the service industry, and genetic engineering has already modified agriculture and pharmaceuticals despite the initial stage of the technological life cycle.

The most significant value is the assessment of the technological factor in the high-tech industry. In a high-tech, progressive industry, for example, producing electronic components, in chemistry, pharmaceuticals or aerospace, technology itself is the main lever that determines an organization's strategy for tomorrow. T. Friedman notes that the experience of high-tech organizations that have not been able to adapt to the rapid market innovations provoked by the above factors can become a warning to other organizations that lack such qualities as leadership, flexibility, imagination in order to adapt, but not because they have no awareness or customer focus, but because the speed of change simply incapacitates them.

In extreme cases, the inability to timely identify the need for a change in technology can lead to significant losses in market positions or make it necessary to complete its activity in a previously profitable business for an organization. However, technology can serve as a fundamental and powerful means by which an organization can gain and maintain a competitive advantage. Past experience indicates that the result of strategic planning of an organization is less sensitive to the specificity of technologies than to its special properties, common to more than a sufficient number of leading industries. Organizations that recognize and own the value of these properties are more likely to be advanced. The change from one technology to another has a more significant impact than, for example, the use of new raw materials, due to the fact that all investments of an organization in the previous technology, in R&D, the main scientific and technical staff of employees, productive potential may become obsolete. For an organization, moving to a new technology is difficult both financially and culturally and politically, because it destabilizes the scientific and technical staff and strong managers' plans to achieve success. Besides, new technology undermines their position of power and influence in an organization. It is also worth noting that when a new technology is radically different from the old one, organizations are usually obliged to unwittingly abandon the area of business where they once were in a dominant position.

Based on the foregoing, the assessment of competitiveness is one of the main economic tasks in the management of industrial sectors. Taking into account the fact that the concept of competitiveness is a synthetic category, it is necessary to build an economic model that will allow to quantitatively assess the competitiveness of an organization based on initial information containing indicators of economic activity and technical characteristics of products. Therefore, to assess competitiveness, it is proposed to apply an economic and mathematical model that will assess the competitiveness of organizations in paired comparisons.

The model proposed here makes it possible to assess the competitiveness of industrial enterprises representing a single industry, since this assessment uses the assumption that the enterprises in question manufacture products of both several types and of a single type.

To assess the competitiveness in one industry, initial data are used that describe the indicators of the economic activity of organizations (which can be taken from accounting and statistical reports or calculated on their basis) and the technical characteristics of products:

1. indicators of financial and economic stability: revenue from sales, production and sales costs, accounts payable, accounts receivable, profit, etc..;

2. indicators of production and sales of products: the volume of R&D expenditures; production volumes (including by groups); sales volumes of products (including by groups); sales volumes of innovative products; export volumes of products (including by group); profit from sales; profit from sales of export products; cost per ruble of sales (the ratio of cost of sales to revenue), etc.;

3. indicators characterizing the condition and efficiency of fixed assets: capital productivity, capital-labor ratio, technical capacity, depreciation factor, expiration date, update rate, payback period of capital investments, etc..;

4. workforce performance indicators: output per employee, profit per 1 employee, share of the increase in turnover due to the increase in output (intensive factor), etc.;

5. indicators of the efficiency of the use of resource potential: the amount of profit per 1 ruble of total resources (capital and labor costs), the amount of turnover per 1 ruble of total resources, etc.;

6. complex calculated indicators characterizing the capacity of implementing an organization's projects: indicator of project feasibility, indicator of project innovation, indicator of the impact of project results (innovative technologies obtained during their implementation) on competitiveness, indicator of the technical level of production of an organization, indicator of the innovative potential of an organization.

Information about the technical characteristics of the manufactured products can be obtained from the product passports.

Thus, the assessment of the organization's competitiveness is based on objective data reflected in the reporting documents and product passports.

As initial data for assessing the competitiveness of products, we present, firstly, vectors describing the economic indicators of an organization $P_1, P_2, ..., P_M$. These vectors consist of the economic indicators of enterprises $P_m = (p_{m1}, p_{m2}, ..., p_{mS})$. Secondly, we take vectors $Q_1, Q_2, ..., Q_M$, where each vector Q_k is a vector with the technical characteristics of the k-th product type:

Note that different types of products may have a different set of technical characteristics, but in the formulation of the model, without losing generality, we can assume that all vectors Q_k have the same dimension.

Further, we will assume that all product names are divided into several types of similar products $Pr_1, Pr_2, ..., Pr_L$.

In doing so, as initial data, it should be indicated which products belong to one type of product, as well as which organizations produce which products.

Using an economic and mathematical model for assessing competitiveness, taking into account the economic indicators of an organization and the technical characteristics of products, the following results are calculated:

1. objective assessment of each product in dimensionless (normalized) units;
2. comparative quantitative characteristics of competitiveness for all pairs of enterprises;
3. quantitative characteristics of the competitiveness of each enterprise;

The described economic and mathematical model uses several auxiliary algorithms:

- algorithm for normalizing vectors with objective characteristics;
- convolution of indicators of objective characteristics of products based on the information feature;
- assessment of competitiveness using a game-theoretic approach.

In addition, the model includes various algorithms and procedures for information processing of initial information.

Algorithm for normalizing vectors with objective product characteristics.

Each product is presented as a vector with the characteristics $Q = (q_1, q_2, ..., q_N)$, where q_n is a separate technical characteristic of a product. This characteristic can be numerical or qualitative; more generally, it is a linguistic variable. In addition, the technical characteristics of products are usually dimensional. Another problem is the fact that the numerical characteristics of a product can be positive when a higher value of a parameter means better production, or negative when a lower value of a parameter means better production.

To assess competitiveness, it is necessary to use the normalized values of technical characteristics. We will evaluate each production indicator using a finite number of points q_{n0} from zero to 5: $q_{n0} \in \{0, 1, 2, 3, 5\}$.

If the initial parameter has a numerical value q_{n0}, then it can be normalized using the following formula (for parameters whose higher value indicates a better indicator):

$$q_{n0} = [5 \cdot (q_n - min) / (max - min)],$$

where *min* and *min* are the minimum and maximum values of the parameter among all products, and square brackets indicate rounding to the nearest integer.

For parameters whose lower value indicates a better indicator, the following version of this formula can be used:

$$q_{n0} = 5 - [5 \cdot (qn - min) / (max - min)].$$

For the case of a linguistic variable or for qualitative characteristics, the expert method for normalization can be used.

Thus, after normalization, we assume that the vectors describing the technical characteristics of a product consist of points of the same scale.

Convolution of objective indicators based on the information principle.

To compare vectors with normalized characteristics, it is necessary to convolve these indicators into a single numerical characteristic. The simplest types of convolution are arithmetic means or geometric means (for positive values). In addition, when performing this convolution, weight factors can be taken into account, which will reflect the importance of each product characteristic.

However, to assess competitiveness, it is also necessary to take into account the importance of each product characteristic in comparison with other (of the same type) product names. To do this, you can use the information principle.

Let there be some set of products of the same type, which we denote by *Pr*. This set will contain the points of the normalized vectors

$$Q_m = (q_{m1}, q_{m2}, ..., q_{mN}), \; m = 1, 2, ..., M.$$

The set of these vectors can be represented as an $M \times N$ matrix, where each row is the indicators of each product, and each column is the values of the indicators in the entire set of products.

For each column numbered n, we define the entropy by the following formula. Assuming that each value of this column is a realization of some random variable o_n, which takes values on $o_n \in \{0, 1, 2, 3, 4, 5\}$, we can estimate the probability for each value of this random variable using the following formulas:

$$p_i = M_i / M,$$

where M_i is the number of times when the exponent n takes the value $I \in \{0, 1, 2, 3, 4, 5\}$.

The entropy of a random variable is calculated using the following formula:

$$H_n = H[o_n] = - \sum p_i \cdot log_2 p_i,$$

where the summation is carried out by i from 0 to 5. As is usual in information theory, we assume that $0 \cdot log_2 0 = 0$.

Thus, for the convolution of the normalized vector with the technical indicators of production, we will use the formula

$$Q^m = \sum_n H_n \cdot q_{mn},$$

where the summation is carried out by n from 1 to N.

Consequently, in the convolution of the indicators, the information content of each technical characteristic is taken into account. The more information each characteristic carries, with the more weight this characteristic is considered in the convolution.

It is possible to use this formula considering the weighting factors

$$Q^m = \sum_n a_n \cdot H_n \cdot q_{mn},$$

where an is the weighting factor for the n-th characteristic.

Assessment of the competitiveness of enterprises using the game-theoretic approach.

Let each enterprise be described by a numerical vector $P = (p_1, p_2, ..., p_G)$. We will assume that the greater the value of each component of this vector, the better the enterprise operates. Let us consider a method for comparing each pair of enterprises – P and Q. Thus, we have two vectors

$$P = (p_1, p_2, ..., p_G) \text{ and } Q = (q_1, q_2, ..., q_G).$$

We form a square matrix $G \times G$ with elements $a_{ij} = p_i - q_j$.

We denote this matrix by $A = \{a_{ij}\}$ and consider it as a payoff matrix in an antagonistic game, that is, the game of two persons with opposite interests. We recall that the matrix game is implemented as follows. Each player has G strategies at their disposal and each player independently chooses one strategy. Suppose that the first player chooses strategy i and the second player chooses strategy j. Then the first player's win (payoff) is defined by the number a_{ij}. Accordingly, the payoff of the second player is equal to a_{ij}.

Let us consider a mixed expansion of this game. As each player's strategies, we will use random variables that take values on the set of strategies. Since the set of strategies is a finite set with G elements, the mixed strategy of the first player is called a random variable ξ, which is given by the vector

$$x = (\xi_1, \xi_2, ..., \xi_G), \; \xi_g \geq 0, \; \xi_1 + \xi_2 + ... + \xi_G = 1.$$

Similarly, the mixed strategy of the second player is a random variable η, which is given by the vector

$$y = (\eta_1, \eta_2, ..., \eta_G), \; \eta_g \geq 0, \; \eta_1 + \eta_2 + ... + \eta_G = 1.$$

The payoff of the first player in a situation of mixed strategies is a number calculated by the formula $H(x, y) = \sum \sum a_{ij} \cdot \xi_i \cdot \eta_j$, where the summation is carried out by i and by j from 1 to G. Accordingly, the payoff of the second player is the number $H(x, y)$.

A pair of mixed strategies x^* and y^* is called an equilibrium situation in the game if for any mixed strategies x and y the inequalities are met:

$$H(x, y^*) \leq H(x^*, y^*) \leq H(x^*, y).$$

In this case, the number $v = H(x^*, y^*)$ is called game value.

According to Von Neumann's theorem, a matrix game always has an equilibrium situation and a game value in mixed strategies.

Assume that enterprise P is more competitive than enterprise Q if the game value is $v > 0$. Moreover, the value v can be used to assess the comparative competitiveness of the enterprise P. If $v < 0$, then enterprise Q is more competitive than P. Its comparative competitiveness is estimated by v.

Algorithm for approximate calculation of equilibrium strategies and game value.

To actually calculate the equilibrium strategies and the game value, linear programming methods or iterative algorithms can be used. Let us consider the Brown-Robinson algorithm, which is also called the effective play method.

Step 1. In the first game, each player chooses an arbitrary pure strategy, for example, the first one.

Step 2. In the k-th game, each player chooses the pure strategy that maximizes his expected payoff against the observed empirical probability distribution of the other player for the previous $(k-1)$ games.

Let the first player choose the i-th strategy ξ^k_i times in the first k games, and the second player uses the j-th strategy η^k_j times. Then in the $(k+1)$ game, the first player will use the $i_{(k+1)}$ strategy defined from the relation

$$v^k = \max \Sigma_j\, a_{ij}\, \eta^k_j = \Sigma_j\, a_{i(k+1)j}\, \eta^k_j,$$

where the maximum is taken by i. Accordingly, the second player will use a $j_{(k+1)}$ strategy defined from the ratio

$$v_k = \min \Sigma_i\, a_{ij}\, \xi^k_j = \Sigma_j\, a_{ij(k+1)}\, \xi^k_j,$$

where the minimum is taken by j.

It is known that $\lim v^k / k = \lim v_k / k = v$, where the limits are taken at $k \to \infty$ and the value of v is the game value. In addition, vectors

$$x^k = (\xi^k_1 / k,\, \xi^k_2 / k,\, ...,\, \xi^k_G / k),\ y^k = (\eta^k_1 / k,\, \eta^k_2 / k,\, ...,\, \eta^k_G / k)$$

will converge to the vectors x^* and y^*, which represent the equilibrium situation.

Let us consider an economic and mathematical model for assessing the competitiveness of organizations in one industry based on the economic indicators of their activities and the technical characteristics of their products. The implementation of this model is carried out in several stages.

Stage 1. Processing of technical indicators of manufactured products.

At this stage, using the described algorithm for normalizing vectors with objective characteristics, we determine the normalized values of technical characteristics. As a result, we get vectors Q_k, whose components are estimated in points $Q_k = \{q_{10}, q_{20}, ..., q_{N0}\}$, $q_{n0} \in \{0, 1, 2, 3, 5\}$.

Stage 2. Convolution of product specifications.

After constructing dimensionless (normalized) vectors of output indicators, it is necessary to perform a convolution operation of these indicators based on the information principle. Using the algorithm described above, it is necessary to calculate the information entropy for each product characteristic based on dividing the set of products into sets of the same type of products

$$Q_m, m = 1, 2, ..., M, Q_m \in Pr,$$

where Γ is the set of the same type of product.

At this stage, it is possible to use weights that will reflect the importance of each indicator.

Stage 3. Creating a single vector for an organization.

Each enterprise under consideration is described by its own vector containing indicators of economic activity. These vectors should be expanded to vectors that contain components for each product name. If an organization does not produce any product name, then the corresponding component of the vector equals to zero, and if it does, the component of the vector is equal to the value of the convoluted product indicator. Thus, we get vectors

$$P_m = (p_{m1}, p_{m2}, ..., p_{mG}), m = 1, 2, ..., M.$$

Next, the resulting vectors must be normalized using an appropriate linear transformation $p_{mg} = \alpha \cdot p_{mg} + \beta$. This transformation should normalize the values of the components that reflect the indicators of an organization's economic activity. In this case, the choice of constants α and β should be carried out so that the corresponding components of the vector satisfy the $p_{mg} \in [0, 1]$ and these indicators should be positive in the sense that a higher value of the indicator means a greater competitiveness of the enterprise. The normalization of the components of the P_m vector describing the production should be a scale normalization that depends on the numerical values. In this case, all components associated with manufactured products should be normalized to the same value $\alpha > 0$, and the value $\beta = 0$.

Stage 4. Generation of matrixes for the game comparison of enterprises, given random factors.

After we have received the normalized vectors describing each enterprise, given the characteristics of the products it produces, it is necessary to generate game matrixes for each pair of enterprises. Let us consider two enterprises P and Q. Thus, we have two vectors $P = (p_1, p_2, ..., p_G)$ and $Q = (q_1, q_2, ..., q_G)$. We form a square matrix $G \times G$, denoted A with elements $a_{ij} = p_i - q_j + F_{ij}$, where F_{ij} are external factors that affect the comparison of two enterprises and which can describe various random factors. In the model implementation, these factors can be described using random variables, for example, distributed normally $F_{ij} \sim N(m, \sigma^2)$.

If these factors can be positive both for the first and the second enterprises, then the mathematical expectation parameter is zero $m = 0$. If these factors are more favorable to the first enterprise, then $m > 0$ should be chosen. In the case when the factors are positive for the second enterprise, then $m < 0$ should be chosen.

Stage 5. Pairwise comparison of enterprises.

Using the obtained game matrices for each enterprise, it is necessary to compare their competitiveness. To do this, using the Brown-Robinson algorithm, we calculate the game value for each pair of enterprise v. Based on the game value (the sign of this value), we order enterprises by competitiveness. The model provides an examination of the competitiveness of enterprises, showing which one is more competitive in comparison with others.

Stage 6. Calculation of the quantitative characteristics of the competitiveness of each enterprise.

At the previous stage, a pairwise comparison of competitiveness based on a game matrix was considered. Moreover, we have the game values for each pair of enterprises.

For each enterprise P, we calculate the quantitative characteristic of competitiveness using the following formula:

$$V(P) = \Sigma \left(v(P,Q_m) + \left| v(P,Q_m) \right| \right) / 2,$$

where the sum is taken for all enterprises, and through $v(P, Q)$ for enterprises Q and P, we consider the price of the game, where enterprise P is analyzed as the first player, and enterprise Q is analyzed as the second one.

Stage 7. Ranking of enterprises according to the characteristics of competitiveness.

After the numerical characteristic of competitiveness $V(P)$ has been calculated for each enterprise, all considered enterprises can be ranked as follows: $P_1, P_2, ..., P_M$ so that

$$V(P_1) \geq V(P_2) \geq ... \geq V(P_M).$$

The proposed method for assessing competitiveness is based on information analysis of indicators of economic activity of organizations and technical characteristics of products. The economic and mathematical model is based on the author's algorithms for normalizing the initial data, the method of informational convolution of vectors containing technical characteristics of products, given the technical characteristics of the same type of products manufactured in a given industry. This approach allows to automatically take into account the informational importance of indicators in convolution. As a result of informational convolution, a single numerical indicator of product characteristics is obtained.

The considered economic and mathematical model for assessing the competitiveness of organizations of one industry is based on a new game-theoretic approach, in which a comparative assessment of the competitiveness of enterprises is made, taking into account the indicators of their economic activity and the technical characteristics of the products. In doing so, along with objective indicators of competitiveness, comparative competitiveness is calculated. Thus, the proposed economic and mathematical model for assessing competitiveness allows covering all the diversity of the concept of competitiveness of high-tech enterprises.

The developed model can be the basis for methods of calculating the competitiveness of industrial enterprises of one industry using information tables containing indicators of economic activity and technical characteristics of products. Through this model an application software can be created to implement this methodology for assessing and examining the competitiveness of high-tech industrial enterprises of one industry.

Thus, when assessing the competitiveness of high-tech industrial enterprises, the following methods are used:

- information model for intelligent processing of indicators of economic activity and technical characteristics of products for subsequent assessment of the competitiveness of high-tech organizations of one industry;
- economic and mathematical model for comparative assessment of the competitiveness of high-tech industrial enterprises of one industry based on information tables describing their economic activities and product characteristics using a game-theoretic approach.
- a comprehensive economic and mathematical model for assessing the competitiveness of high-tech industrial enterprises of one industry based on the generated information table of indicators of economic activity and technical characteristics. From the use of this model, the following are calculated: an objective assessment of each product in dimensionless (normalized) units; comparative quantitative characteristics of competitiveness for all pairs of high-tech industrial enterprises; quantitative characteristics of the competitiveness of each such enterprise.

We emphasize once again that the economic and mathematical model is the basis for creating working methods for assessing the competitiveness of high-tech enterprises of one industry sector, taking into account the indicators of their economic activity and technical characteristics of products, as well as for the implementation of an information and analytical system for assessing and examining the competitiveness of these enterprises.

It is also worth noting that the gaining competitive advantage by high-tech enterprises is positively influenced by activities related to the implementation of innovations: R&D, implementation of technological innovations, training of employees, release of new products. Accordingly, an increase in the average age of equipment reduces the competitiveness of an organization.

5.3. ASSESSMENT OF THE ECONOMIC SUSTAINABILITY OF THE ORGANIZATION IN THE PROCESS OF UPDATING PRODUCTS

A progressive market economy defines economic sustainability as one of the main requirements for the continuous functioning and development of high-tech organizations. A systematic approach to the development of particularly optimal ways to manage economic sustainability, considering the experience of advanced countries, as well as the reasons for the influence of external conditions, can be the basis for a comprehensive long-term plan for the economic development of high-tech organizations.

Economic sustainability can be viewed as a concept in economics when an organization, despite the negative impact of external and internal conditions, can maintain a stable position, as well as function and develop productively.

The functioning of any process always strives for consistency and thus limits the development of the process and at the same time finances it. In turn, development complicates individual stages of functioning; however, in general it creates opportunities for its sustainability.

Initially, the most important is the mission of an organization, followed by the goals of functioning and development. Each organization aims to maximize the value of its funds, while maintaining the indicators of asset liquidity and production results. An additional goal is to achieve maximum profit for a specific period of time. The constant increase in profit indicators over a long period of time leads to an increase in the cost of an organization.

The financial plan (also the totality of funds in an organization – Cash flow) is a well-known method for identifying profits from the execution of projects and the foundation of basic methods for analyzing planned projects. It characterizes the efficiency of financial management and is used in most common methods of forecasting the efficiency and financial results of the production and investment activities of an organization.

It is important to organize the financial planning so that there is optimal balance of funds for a certain period of time project on the current, investment and financial activities of an organization, which allows identifying the need for capital, develop strategic, tactical and operational plans to obtain financial indicators and establish the productivity of the workplan.

As already noted, an increase in the value of funds of any organization (for example, a high-tech corporation) is possible if you get the maximum profit for a long time. It follows that the economic sustainability of an organization can be defined as such a state of indicators of production and the economy of an organization, where their values make it possible to obtain the necessary profit in a certain period.

When conducting an economic analysis of an organization, many predominantly optimal economic performance indicators are calculated, which must be reduced to a single indicator.

The applied approaches to assessing economic sustainability are based on a systematic approach to the functioning of organizations and the calculation of a single sustainability indicator.

There are different views on these issues. For example, A.A. Zotov proposes a technology for calculating a single indicator of economic sustainability, but does not classify indicators into groups based on common characteristics. Thus, it is difficult to assess the economic sustainability of an organization due to the need to instantly identify the degree of indicators with different characteristics.

Another researcher, S. N. Anokhin, suggests considering economic sustainability as sustainability: management, production, financial, social, business activity and profitability.

It is important to note that the methodology proposed by the author does not consider rank coefficients, which results in the the impossibility of taking into account the value of each stability coefficient in the formation of the integral coefficient.

Another scientist, I.V. Bryantseva assesses the economic sustainability of an organization in comparison with the selected complex of similar organizations, but does not take into account the movement of the integral indicator within only one organization. The author focuses on two components of the economic sustainability of an organization: production and technical, as well as financial and economic. Production and technical sustainability consists of the sustainability of technical and technological, personnel, organization of production, while financial and economic sustainability implies financial sustainability, sustainability of business activity, sustainability of the level of profitability.

The authors A. Voronov and S. Rubanov, when assessing technical stability, take into account the dynamics of indicators that reveal the number of increases (decreases) over a certain period of time and whose transformations can have a positive and negative effect. For all indicators, the coefficient of sustainability of the further development of an organization in the considered stage is determined. Researchers highlight such indicators of economic sustainability as: the level of produced and sold goods and works, the financial status of projects, scientific and technological progress, properties of commodity and the proposed works, expenses for each unit of goods produced and provided works of raw materials, labor and monetary resources. The main conclusion of the proposed methodology is the calculation of the dynamics of indicators without taking into account specific values, which probably can lead to incorrect results regarding the level of economic sustainability of an organization. It is also important to note that in some places there are no rank coefficients and any indicators in the calculations have the same values.

A large number of existing technologies assess the economic sustainability of organizations, based on full assessment of their operation and the calculation of a complex factor of economic sustainability.

Any single factor showing the economic activity of an organization can affect the complex factor of the economic sustainability of the organization itself. The degree of influence of a single factor on a complex factor is determined using ranks.

To simplify the ratio of the rank coefficients of the influence of a single factor on the summarizing ones, it is necessary to create a combination of private indicators for the same properties. A special solution in the management of economic order at different levels (macro, medium, micro) is the concept of sustainable development, enabling a holistic approach to the consideration of multiple and dual problems of economic development, leading to a productive growth plan, at the same time taking into account the differences between mechanisms, given the coherence of cooperation and coherence for mutual benefit.

Productive economic development of organizations is a difficult, multifaceted task and requires research on this issue.

The creation of the necessary mechanisms and the choice of technologies for the sustainable development of organizations are predetermined by the final signs of the degree of its economic sustainability (hereinafter referred to as ES), which shows an organization's ability to be consistent with changing balance as a result of growth, productive use of planned and economic growth opportunities, thereby obtaining the results of attitudinal growth factors in the case of the influence of external and internal circumstances.

The results of a study by a number of scientists of modern methods of assessing an organization's ES show that there is no systematic solution to this problem yet. It is possible to group the given assessment methods as follows:

1. by the number of values in the sustainability system-one-factor and multi-factor methods;
2. by the nature and methods of factors used in the sustainability analysis (applied and changing factors);
3. by the leading order model for analyzing the degree of sustainability -methods based on the speed of grouping factors and on comparing their exact values with legitimate systems of complex analysis of sustainability.

It should be emphasized that the technology for analyzing an organization's ES should take into account the movement of the organization's growth in the main types of work, as well as maintaining the balance of factors in the components of ES, including industrial, innovative, material, organizational and administrative and promoting stability on the market.

Based on the completed study of the ES, we will highlight the main aspects for the formulation of the analysis methodology. The technology for analyzing an organization's ES as a result of the observed growth necessarily includes the following components: goals and objectives; subjects of analysis; analysis factors mechanism; the consistency of the analysis; methods for calculating factors; information for analysis; mechanical techniques for performing analysis; preparation of analysis (identification of persons for analysis, regularity of analysis, etc.). In our opinion, the analysis of the sustainability of an organization, in particular a high-tech corporation, necessarily takes as a basis such units as: a complex unit of analysis containing the factors of elements of the main aspects of the economic growth of an organization; temporary growth qualities of the main aspects; growth oriented analyzes of main aspects; legal units for specific qualities. Based on the results of ES analysis methods, we present the technology of complex analysis of an organization's sustainability.

Within the framework of this technology, one of the main stages has been analyzed – the creation of a mechanism of factors that accurately reflect the degree of ES of an organization due to its growth. An accurate analysis of the ES should be based on criteria that show maximum stability of growth in the most important aspects; however, it always reveals constant factors, which is a disadvantage. To solve this problem, it is necessary to move to the intensive development of indicators (growth rate), upon condition of the regulatory movement and norms of important aspects of the development of an organization itself (for example, for a production organization – release of goods, introduction of innovations, coordination moments, funds, product promotion). Taking into account the above aspects and based on the constructed technology for analyzing ES as a result of the growth of an organization, it is necessary to summarize the factors reflecting all possible forms of its financial and economic activity and the components of ES, and distribute their investment according to the proposed elements of ES.

It is recommended that all factors be considered when analyzing the elements of ES. For example, in order to analyze the stability of an organization's funds, factors reflecting the state of funds, timely fulfillment of monetary obligations, stability, profitability, activity, and an organization's interest for investors should be considered. In this complex of factors, the main conditional factors can be identified, including indicators of distribution of net profit, turnover of assets in money, cash leverage, return on capital, personal and invested funds, resource return, cash of an organization, dividend and creditor obligations, etc.

To create a changing scheme for analyzing an organization's ES in the course of its growth, it is necessary to calculate the rate of development of single factors of sustainability elements. However, for any factor, it is advisable to determine the legal force of the change.

The need to solve the problems of economic sustainability of organizations, to guarantee their greater adaptability to the market implies the need to study and identify a system of criteria and indicators that allow, firstly, to assess the level of sustainability and dynamics, and secondly, to identify factors that can reduce their negative impact on economic sustainability.

At the same time, Russian high-tech corporations face the primary and urgent task to ensure the competitiveness of their products in the local and world markets.

The competitive advantage of any high-tech product consists of the impact of many reasons that appear during the distribution of this product in the markets. The real competitive advantage of a product is its sale.

It is possible to take into account the described reasons within the limits of high-tech production, for example, helicopter construction, if one arrives at the result under conditions of inaccuracy. Inaccuracy is manifested in the results of market activity, in the distribution of goods and their competitiveness.

The mathematical scheme for considering internal and external causes that appear during the launch of updated products on the market can be shown as follows.

It is necessary to analyze the set of products that compete with products on the local market:

$$N_i, \quad i = 1, 2, \ldots, n \, .$$

In the same way, a set of goods that competes with a product in the foreign market is analyzed:

$$M_j, \quad j = 1, 2, \ldots, m.$$

The complex factor of competitive advantage Q is calculated taking into account the local and foreign markets. Then the factor of competitive advantage will be significantly determined by the cost of the goods and the cost of the competing goods. Below is the equality that expresses the complex factor of competitive advantage:

$$Q = F_{in}(C_{in}, C_{in}^1, C_{in}^2, \ldots, C_{in}^n) + F_w(C_w, C_w^1, C_w^2, \ldots, C_w^m),$$

where F_{in} is the functionality of the competitive advantage in the local market;

F_w is the functionality of a competitive advantage in the foreign market;

C_{in} is the price of the product in the local market;

C_{in}^i is the price of competing goods in the local market;

C_w is the price of goods on the foreign market;

C_w^j is the price of competing goods on the foreign market.

Here are three types of indicators of updated products that affect the competitive advantage of the product:

- indicators reflecting the techno-economic qualities of the products;
- economic indicators (cost);
- arbitrary indicators.

Economic factors are initially included in the price of goods in the local and foreign markets.

The techno-economic qualities of a product affect the complex factor of competitive advantage. With regard to the factors that reveal the techno-economic qualities of a product, let us consider the values that consider these qualities:

a_{in} – the value that takes into account the techno-economic qualities of the product on the local market;

a_w – the value that takes into account the techno-economic qualities of a product that is competitive in the foreign market;

a_{in}^i – the value that takes into account the techno-economic qualities of a product that is competitive in the local market;

a_w^j – the value that takes into account the techno-economic qualities of a product that is competitive in the foreign market.

The greater the parameter value, taking into account techno-economic quality of goods, the more suitable it is for the parameter of the final factor of its competitive advantage.

So, arbitrary internal and external indicators affect the final factor of the competitive advantage of the product produced using complex technologies. In relation to such indicators, we will accept an arbitrary parameter of the transformation of a competitive advantage under the influence of arbitrary indicators:

ρ_{in} – the transformation rate of the competitive advantage of a product under the influence of arbitrary indicators in the local market;

ρ_w – the transformation rate of the competitive advantage of a product under the influence of arbitrary indicators in the foreign market;

ρ_{in}^i – the transformation rate of the competitive advantage of a product that has competitiveness under the influence of arbitrary indicators in the local market;

ρ_w^j – the transformation rate of the competitive advantage of a product that has competitiveness under the influence of arbitrary indicators in the foreign market.

The coefficient of transformation of the competitive advantage of a product under the influence of arbitrary internal and external indicators affects the final parameter of the complex factor of the competitive advantage of goods produced using complex technologies thus: growth factors and increases them.

Therefore, after determining the indicators that affect the competitive advantage of the product, the complex factor of competitive advantage will be calculated as follows:

$$Q = F_{in}[q_{in}(C_{in}, a_{in}, \rho_{in}), q_{in}^1(C_{in}^1, a_{in}^1, \rho_{in}^1), \ldots, q_{in}^n(C_{in}^n, a_{in}^n, \rho_{in}^n)] +$$

$$+ F_w[q_w(C_w, a_w, \rho_w), q_w^1(C_w^1, a_w^1, \rho_w^1), \ldots, q_w^n(C_w^m, a_w^m, \rho_w^m)] \ ,$$

where $q_{in}(C_{in}, a_{in}, \rho_{in})$ – the function of a single competitive advantage of a product in the local market;

$q_w(C_w, a_w, \rho_w)$ – the function of a single competitive advantage of a product in the foreign market;

$q_{in}^i(C_{in}^i, a_{in}^i, \rho_{in}^i)$ – the function of a single competitive advantage of a competitive product in the local market;

$q_w^j(C_w^j, a_w^j, \rho_w^j)$ – the function of a single competitive advantage of a competitive product in the foreign market.

The effect of the transformation rate of the competitive advantage of a product under the influence of arbitrary internal and external indicators on the parameter of the functions of a single competitive advantage is as follows: increasing the factor ρ increases the parameter of the single competitive advantage function. In the same manner, an increase in the factor a, which takes into account the techno-economic qualities of a product, leads to an increase in the parameter of the function of a single competitive advantage of the product.

Next, we analyze the impact of the parameters of the functions of a single competitive advantage on the final parameter of the complex factor of competitive advantage. To analyze the numerical impact of a given value, it is important to parse the appropriate derivatives.

The derivative $\dfrac{\partial Q}{\partial q_{in}}$ equals

$$\frac{\partial Q}{\partial q_{in}} = \frac{\partial F_{in}[q_{in}(C_{in}, a_{in}, \rho_{in}), q_{in}^1(C_{in}^1, a_{in}^1, \rho_{in}^1), \ldots, q_{in}^n(C_{in}^n, a_{in}^n, \rho_{in}^n)]}{\partial q_{in}} > 0.$$

Which suggests that with an increase in the parameter of the function of a single competitive advantage of a high-tech product in the local market, the parameter of the complex factor of competitive advantage increases.

The derivative $\dfrac{\partial Q}{\partial q_w}$ equals

$$\frac{\partial Q}{\partial q_w} = \frac{\partial F_w[q_w(C_w,a_w,\rho_w),q_w^1(C_w^1,a_w^1,\rho_w^1),\ldots,q_w^m(C_w^m,a_w^m,\rho_w^m)]}{\partial q_w} > 0.$$

This means that with an increase in the parameter of the single competitive advantage function of a high-tech product in the foreign market, the parameter of the complex competitive advantage factor increases.

The derivative $\dfrac{\partial Q}{\partial q_{in}^i}$ equals

$$\frac{\partial Q}{\partial q_{in}^i} = \frac{\partial F_{in}[q_{in}(C_{in},a_{in},\rho_{in}),q_{in}^1(C_{in}^1,a_{in}^1,\rho_{in}^1),\ldots,q_{in}^n(C_{in}^n,a_{in}^n,\rho_{in}^n)]}{\partial q_{in}^i} < 0.$$

That is, with an increase in the parameter of the function of a single competitive advantage of a product in the local market, the parameter of the complex factor of competitive advantage decreases.

The derivative $\dfrac{\partial Q}{\partial q_w^j}$ equals

$$\frac{\partial Q}{\partial q_w^j} = \frac{\partial F_w[q_w(C_w,a_w,\rho_w),q_w^1(C_w^1,a_w^1,\rho_w^1),\ldots,q_w^m(C_w^m,a_w^m,\rho_w^m)]}{\partial q_w^j} < 0.$$

Hence, with an increase in the parameter of the function of a single competitive advantage of a competitive product in the foreign market, the parameter of the complex factor of competitive advantage decreases.

The specificity of high-tech products is that a small number of the same types of products are sold in the domestic and foreign markets. It is worth noting that in high-tech industries, the threshold for entering the world and domestic markets is very high. It follows that the methodology proposed in this study for assessing the competitiveness of products, taking into account the factors arising from the sale on the world and domestic markets, can provide operational tracking of the dynamics of product competitiveness.

In modern conditions of world economic development, the use of these methods is extremely important, since the life cycle of innovations, the cycle of development and production are reduced. For this reason, even at the stage of the idea of developing an innovative product, it is necessary to take into account whether the project will have commercial success and whether it will be able to create a new consumer market.

Previously, various opinions of Russian authors about the economic sustainability of an organization were considered, from which it is clear that there is no generally accepted formula for its calculation, just as there is no generally accepted definition of this concept. Scientists apply various complexes of factors and systematize them in different ways.

Since the internal indicators that affect the ES of a high-tech organization are its current, investment and material activity, the components of the complex factor of its economic stability consist of factors of current, investment and material stability.

When analyzing the economic stability of organizations, information about internal and external working conditions is used. The assessment of external conditions, including their control, in comparison with the internal conditions of an organization is much more difficult. The use of internal information in the analysis of the economic stability of organizations increases the validity of the analysis.

The form and technology of analysis of organizations' ES in the period of their growth, based on the complex significance of analysis, the study of the growth of the main aspects of work, the legitimacy of the movement of factors in the analysis of elements of an organization's ES, dynamic comparability and interdependence of factors make it possible to identify the best aspects of an organization's growth, and also are the basis for creating tools for its stable economic growth.

CHAPTER 6. PRODUCT RENEWAL AS A MECHANISM FOR MAINTAINING GLOBAL COMPETITIVE LEADERSHIP

6.1. VOLUME INCREASE OF OWN PROCESSING WHEN UPDATING PRODUCTS AS A SOURCE OF CREATING COMPETITIVE ADVANTAGES FOR AN ORGANIZATION

In modern conditions, advanced development of an organization, consisting in ensuring a dominant position in the market, is a necessary sign of the successful and effective operation of high-tech holdings and science-intensive industries. Product updating is an integral part of managing and maintaining advanced development. Advanced development refers to the management of research, development, innovative, production, organizational and other competencies of an organization, which determines the formation of an organization's potential sufficient to create a completely new market for high-tech products or to gain a significant share in the existing market, as well as maintain a leading position in the market for a long period.

Usually, with such a complex of interrelated activities, a single organization within a holding cannot perform them completely independently without interaction with other enterprises or organizations. Practical development for the creation of complex high-tech products is mainly carried out by several design bureaus. The production of complex equipment can be carried out by different enterprises in cooperation with each other. When managing holding structures, it is necessary to take into account mutual production and technical ties and introduce them into the design, development and production of innovative technologies, which will increase the competitiveness of manufactured products, thereby increasing the competitiveness of a holding itself. When creating a holding structure, it is necessary to address the issue of comprehensive integration of the competencies of member-organizations. The synergistic effect of such integration makes it possible to quickly implement and use strategic innovations that allow effective competition in domestic and foreign markets. Active use and development of a holding's unique competencies can create a market of fundamentally new products and technologies and create new market segments based on them.

It is necessary that the set of the main projected long-term goals and objectives of an organization's advanced development occupy a central place in its development strategy and be consistent in terms of resources and time frame.

The indicators that determine the sustainable technical development of an organization characterize the existing technical level of production and the ability to reformat its activities to manufacture advanced development products ensuring a leading position for a long period of time.

A key factor that determines the possibility of advanced development of an organization is its ability to reach a new level of industry positioning, which in turn is determined by the ability to develop and implement advanced technologies inside the industry or in related industries. The following levels of industry positioning can be distinguished: narrow industry positioning (competitiveness in one area); broad cross-industry positioning (competitiveness in several industries); interdisciplinary positioning (competitiveness in many areas). For the analyzed organization, it is necessary to form a bank of unique competence that will allow it to reach new levels of industry positioning.

A special case of creating high competitive advantages of products is the process of product update, which is designed to ensure the maintenance of a company's leading position in the market, as well as reaching a new level of industry positioning.

In scientific literature and in practice, two methods are proposed for developing an updating.

The first method considers the increase in the economic effect from the product updating as a criterion, in which (given the standard efficiency coefficient) the maximum possible production period is calculated; the second method – the product updating period is established based on the existing trends of its obsolescence.

Stimulating the product update and the introduction of new types of products presupposes the creation of conditions that ensure a high material interest of manufacturers and consumers in the modern replacement of outdated equipment. It is assumed that new types of products should be more profitable for manufacturers and consumers.

For economically feasible management of the process of updating products at the stage of their development in serial production, it is necessary to implement a number of practical proposals that should increase the incentive of enterprises to release products, in particular, cost recovery of increased production.

Interest in updating products is also characterized by the indicator of specific net profit, which reflects all the main aspects of the economic activities of organizations and the industry as a whole. This indicator is also supported by the fact that the profit used for its calculation gives the measurement of the technical level the character of assessing profitability, and the number of employees – the character of assessing labor productivity.

The period of product update should be within provisional borders, the lower limit of which is the time required to release the minimum specified volume of new products, and the upper limit is the time required to produce its maximum volume, taking into account obsolescence, determined by the growth rate of the efficiency of social production.

The stimulating role of price is of particular importance in managing product update. In the production of developed products, two or more price levels can be approved with fixed terms of their introduction (step prices). They should be established to stimulate the production update and its most rational use. The price system can be limited to three steps.

The first step is a stimulating price, which creates favorable conditions for the production and development of designed products to replace manufactured types of products. The price should be set closer to the upper limit.

The second step is the average price, which provides a company with the necessary profit to establish funds.

The third step is the penalty price. This step reduces profitability due to the loss of product novelty.

As can be seen from the graph (Fig. 6.1), the expenditure of a product manufacturer replacing the manufactured types of products is not the same throughout the entire period of its manufacture. A step price for updated products is adopted. For the first two years, the contract (temporary) price is of the greatest importance, which is due to additional expenditure to achieve the projected quality indicators, reflected in the price through quality coefficients. In the initial period, the cost is most important; then, as the serial production increases and the release of updated products is mastered, the cost reaches the lowest value by the sixth year of serial production. At the third step, the price is set given the discount, which takes into account the obsolescence of the product and the cost rises. Demand and profits for these products are declining, which prompts the manufacturer to update them.

To solve the problem of industrial development of products at enterprises, special development capacities should be devoted. In this case, the transition to new types of products will not adversely affect the production of basic products.

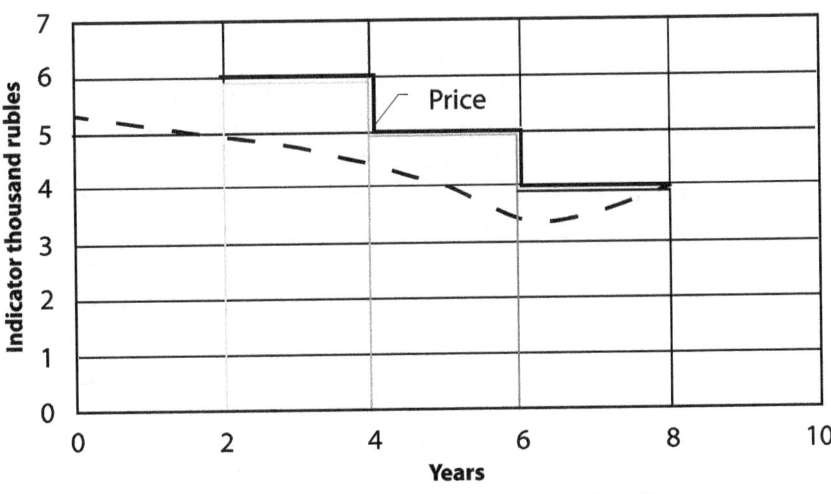

Figure: 6.1. Changes in price, cost and profit

Finally, in order to accelerate the innovative development and development of new products, it is necessary to improve or revise the management system of an organization, so that the introduction and development of new technologies becomes a common phenomenon for employees.

Concerning the development of new products, it should be noted that when creating high-tech products in order to achieve high indicators of competitiveness and to dominate the market, designers put the most modern materials, components, units, devices and systems in the design of the product. Specifying production processes are carried out by the developer enterprise: the creation of housings, the production and assembly of the main product, its customization, adjustment, testing, and ultimately implementation. Moreover, even in the case of designing a product for a given cost, its market entry does not provide a sufficient period of dominance, both in terms of technical indicators, and in terms of economic and price indicators. This arises from the fact that the market, as a rule, operates companies offering similar products, and competitors of manufacturers seek to increase their competitive advantages. On the other hand, the price of products begins to increase as a result of the negative impact of external factors (inflation, currency exchange rate changes, foreign economic sanctions, etc.) both on the cost of materials and components, and on the cost of the finished product itself. In these conditions an organization's management faces two challenges: to work with suppliers to optimize the cost of supplied parts, units and assemblies, so that their value remains stable with dynamically changing factors; simultaneously perform similar activities to reduce cost, taking into account the emerging factors of the forced price increase within the enterprise itself.

The lack of an effective solution to these problems leads to a change in an organization's income. In this case, in order to reduce the cost of finished products in a short time, outsourcing services for the production of individual components of the product can be involved as a temporary solution. The saved financial resources should be directed to the own pre-production, which will reduce the cost of manufactured parts, adjustment and customization works and get eliminate dependence on an outsourcing company. It is necessary in view of the fact that after a certain time (usually, a year or two) an organization loses its competence to produce an outsourced component, and the outsourcing company begins to raise the cost of its services, realizing that the company is forced to purchase products in order to fulfill its obligations to their customers. In such a situation, managers need to determine the optimal amount of work that can be outsourced and the "corridor" of changes in prices for the products of subcontractors in order to level the risk of bankruptcy, which is not profitable for either the organization itself, nor its subcontractors and partners. This is due to the fact that with a reduction in the volume of its own processing, an organization will not have enough funds to finance development and modernization of production, create competitive advantages of already manufactured products, and develop new products. Thus, it faces the task of determining the optimal volume of its own processing, providing a sufficient revenue part for carrying out measures to develop production, create competitive advantages of products or update the product range.

A significant amount of own processing is one of the ways to ensure economic sustainability. Outsourcing a particular stage of the production process in most cases leads to an increase in variable costs. When assessing the dynamics of fixed costs when outsourcing part of the work, it is necessary to take into account the reduction in existing costs and the emergence of additional ones, for example, for the remuneration of new employees appointed to interact with the outsourcing service provider.

Let us describe the process of finding the optimal balance between an organization's own processing and outsourcing of part of the work using the economic indicator $EBIT_{\%}$ of operating profit:

$$EBIT_{\%} = \frac{\sum_{i=1}^{N}(p_i - v_i)Q_i - FC}{\sum_{i=1}^{N} p_i Q_i},$$

where p_i is the cost of a unit of product type i; v_i are variable production costs per unit of product type i; FC are fixed production costs; Q_i is the output volume of the product type i; N is the number of types of a product.

Consider the change in the $EBIT_\%$ indicator due to the outsourcing of part of the work. According to the assumptions above, fixed costs are reduced when part of the work is outsourced. The rate of reduction of fixed costs of an organization decreases with an increase in the volume of work outsourced, variable costs increase in proportion to the volume of work outsourced. Suppose that the fixed costs of an organization in the absence of outsourcing are $FC=3700$. The volume of output of each type of product, the unit cost of each type of product, fixed costs and variable production costs per unit of product and the calculated values of the $EBIT_\%$ indicator are presented in table. 6.1. In the calculated example, 10 to 80% of the work is consistently outsourced. The graph of changes in the $EBIT_\%$ indicator value is shown in Figure 6.2.

Table 6.1. Initial data for calculating the operating profit of an organization with different volumes of its own processing

Type of work I	Product component cost p_i	Variable costs for own processing in outsourcing v_i	Fixed costs for outsourcing i works	Required scope of work of type iQ_i	Operating profit $EBIT_\%$
1	20	10\12	3500	50	0,264
2	30	10\12	3320	45	0,269
3	27	12\15	3150	30	0,276
4	26	10\13	2990	55	0,283
5	35	13\14	2840	45	0,282
6	30	9\11	2700	40	0,291
7	30	15\17	2570	50	0,296
8	25	10\15	2410	45	0,293
9	20	11\12	2480	30	0,287
10	35	15\16	2330	45	0,275

Operating profit under the conditions of the example presented reaches its maximum value at 40% of its own processing, which is quite consistent with the practice of leaders of world industrial production. Thus, the organization maximizes its profits and receives a resource for long-term development.

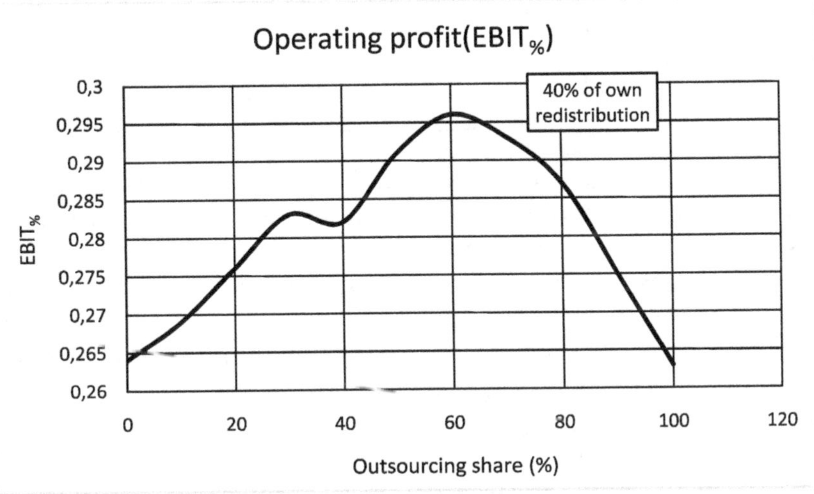

Figure: 6.2. Change in operating profit upon change share
of outsourced work

The economic sustainability of an organization can be assessed by the level of impact of the *DOL* operating leverage, whose calculation formula in this case will be as follows:

$$DOL = \frac{\sum_{i=1}^{N}(p_i - v_i)Q_i}{\sum_{i=1}^{N}(p_i - v_i)Q_i - FC}.$$

The economic meaning of the value is as follows: the greater the share of fixed costs in the structure of production costs, the higher the level of impact of operational leverage and, consequently, the higher the production risk.

In the context of the previous example, consider changes in the level of impact of production leverage when outsourcing part of the work. Based on the data presented in Table 6.1, we calculate the level of impact of the operating leverage when outsourcing various volumes of work and when adding new types of products to the output (7–10). The calculation results are presented in table. 6.2.

Table 6.2. Calculation of the operating leverage DOL

Indicator	Outsourced work types				
	1–2	1–4	1–6	1–8	1–10
DOL	5,3	3,3	2,62	3,45	4,07

The DOL ratio is directly related to the level of operating risk: the greater the level of impact of operating leverage, the greater the risk. In this regard, a decrease in the operating leverage also indicates an increase in the economic sustainability of an organization, and its increase indicates a decrease in economic sustainability. The calculation shows that outsourcing more than 60% of the work leads to a loss of economic sustainability. The results obtained can be confirmed by statistics from world industrial leaders.

Figures 6.3 and 6.4 present data on the development of cooperation between the world's leading aircraft manufacturers Airbus and Boeing. Clearly, cooperation is becoming an important factor in the development of modern global corporations. Analysis shows that the enhanced cooperation is a forced measure of a large company aimed

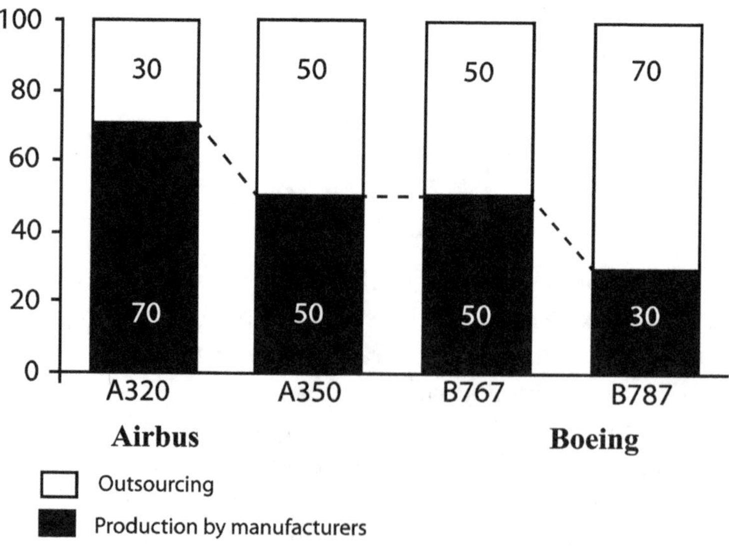

Figure: 6.3. Share of outsourcing and production cooperation in Airbus and Boeing

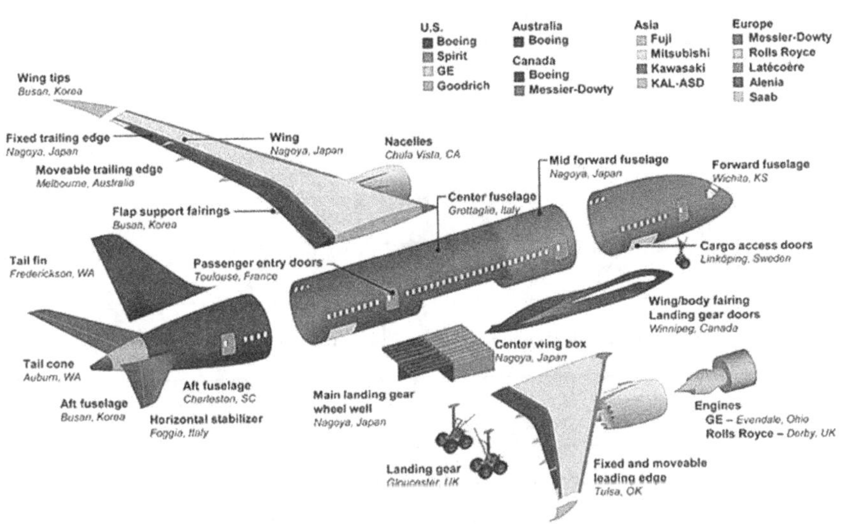

Figure: 6.4. International cooperation in the production of Boeing 737[21]

at maintaining its competitiveness. This is largely due to the fact that the development and production of final high-tech products requires a large number of components (machines, units, parts) that are produced by enterprises of other industries.

Their development and production requires other competencies and equipment that enable the design of quality products in the relevant area.

According to the data presented in Fig. 6.3, the average in-house production is 40% for Boeing and 60% for Airbus.

Among the leading Russian industrial enterprises, the volume of their own processing is in the range of 30-60%.

The decision to outsource part of the work for a short time in order to optimize the cost of production can be made on the basis of the following algorithm (Figure 6.5).

The decision to use outsourcing is a difficult task affecting the strategic interests of the parent organization related to ensuring its sustainable economic development. The practice of world industrial

---

21 Source: Chesbrough, H. Open innovation. Creation of profitable technologies [Text]: monograph / G. Chesbro. – M.: Pokolenie, 2007.–336 p.

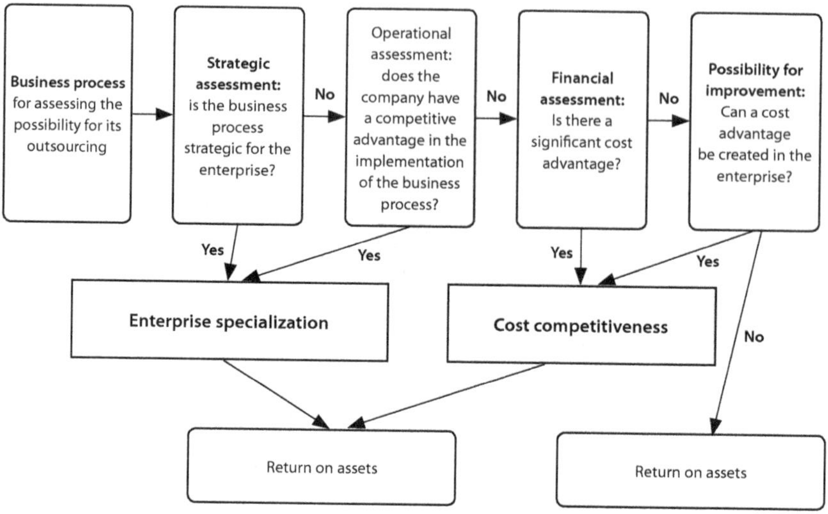

Figure 6.5. Algorithm for deciding on work outsourcing

leaders indicates the need to preserve at least 30–40% of their own processing of key products. The starting point for deciding whether to manufacture in-house or to accept third-party services is to analyze your own production management processes as well as product implementation based on strategic perspectives. This issue is related to the building of digital twins of products and organizations, forecasting changes in the production cost due to the dynamics of market prices for materials and components and planning the volume of own processing, which provides a profitable part for the sustainable development of an organization.

6.2. APPLICATION OF MODERN DIGITAL TECHNOLOGIES IN THE CREATION OF COMPETITIVE PRODUCTS

Creating an advanced product image is one of the most important and complex tasks, which are currently being effectively solved through the use of intelligent design systems. The image of a product gives an idea of its techno-economic indicators. Achievement of these indicators is determined by such factors as the competence of designers, the availability of a modern product and manufacturing base, the ability to exchange data with the international information space.

Thanks to modern technologies and programs, organizations have access to ready-made design documentation after creating a digital twin of a product (a virtual image with all the physical and techno-economic properties of the planned product and allowing its virtual testing). Thus, the development of design documentation can be carried out using computer programs (in particular, the NX Check-Mate program from the family of PLM applications of Siemens).

For example, the use of this NX-system and Teamcenter at Mil Moscow Helicopter Plant (Russian Helicopters Holding) led to the following positive results:

- full transition of an organization to digital creation of new and modernized products;
- reduction of the start-up period for prototypes production;
- significant improvement in the quality of manufactured equipment;
- original status of design documentation for electronic 3D models;
- electronic approval and approval of design documentation in the Teamcenter environment, which currently has more than 600 registered users. Among other things, Teamcenter also stores scanned paper documentation, so that drawings and other documentation are available to users at any time;
- improving the quality of the created 3D models;
- reduction of terms of electronic approval of design documentation by 10-15%;
- technical preparation of production based on 3D models of parts and units of helicopters;
- reducing the number of design errors and flaws;
- reducing the cost of product improvements when putting into production[22].

---

22 Mil Moscow Helicopter Plant uses Siemens PLM Software digital technologies to design and manufacture helicopter technology.

Consider the methods used to create a digital twin of a product at various stages of its life cycle:

- digital design method at the stage of product development;
- method of building a digital twin of an organization at the stages of pre-production and production.

In addition, approaches to the creation of a digital twin of products are somewhat different in various PLM systems (software systems for product lifecycle management, including the creation of a digital twin) through the use of special methods (for example, economic analysis of the effectiveness of the development of a digital twin).

The most important method of creating a digital twin is digital design and modeling, which can be implemented using such tools as PLM systems (i.e., product lifecycle management systems) and other computer-aided design and modeling systems (CAD systems).

The digital design tool is the creation of a digital model of a product in the PLM system (or other CAD systems), which is subsequently transferred to the command center in 3D format. In a matter of seconds the application is able to create thousands of virtual product variants, which exactly matches the developed product. At the same time, to identify potential problems and necessary improvements, the big data processing technology and product and manufacturing information (PMI) contained in the modules (tolerances, fit, connection between parts and units), as well as a basic description of the technological process are used. This tool has been tested many times and is actively used in practice, for example, when creating a digital twin of products.

Such a digital representation of the product remains relevant throughout the entire life cycle of a product and may contain:

- digital model of the product;
- material specification;
- management and data maintenance of the product;
- information about the product's behavior under various conditions.

The most complete and complex assessment of the technical qualities of, for example, a helicopter is the conduct of flight research in flying laboratories (FL) with variable characteristics of stability, controllability and control systems. Conducting such a study makes it possible to form recommendations on the structure and laws of automated helicopter control, which ensure optimal controllability, specified piloting accuracy and flight safety. This type of testing is extremely expensive. Conducting a test on a digital twin of the developed helicopter through digital organization is a way to minimize the cost of its preparation and implementation. Virtual testing of helicopter equipment, in addition to minimizing costs, also reduces the time to bring the helicopter to market, since it takes much more time to prepare a full-scale test in a flight laboratory than to prepare a similar virtual test.

As a result of many virtual automatically conducted tests, a certain model is created that allows programming the destruction of helicopter parts at the junctions in various emergency situations in such a way that the level of passive safety established by the requirements is achieved, in which nonlinear deformation and destruction of the body will ensure the survival and minimum possible injury to pilots and passengers, as well as cargo damage. The high detail of the digital twin of a model programs the behavior of each of its elements in operation or accident conditions.

Thus, a digital twin has a high level of utility, provided that the cost of developing or purchasing the system necessary for building a digital twin, used for its design and modeling (PLM systems, etc.), is lower than the cost of creating a product sample and its full-scale testing.

As the implementation of a PLM system is very expensive, it is recommended to pre-analyze the economic efficiency that an organization can obtain when implementing modern PLM systems. Figure 6.6 shows a five-level model of digitalization of production.

The effectiveness of a digital twin of a product is achieved due to the synergistic effect in a single digital system arising from the integration of high-precision physical and mathematical models, artificial intelligence, computer training and software training analytics,

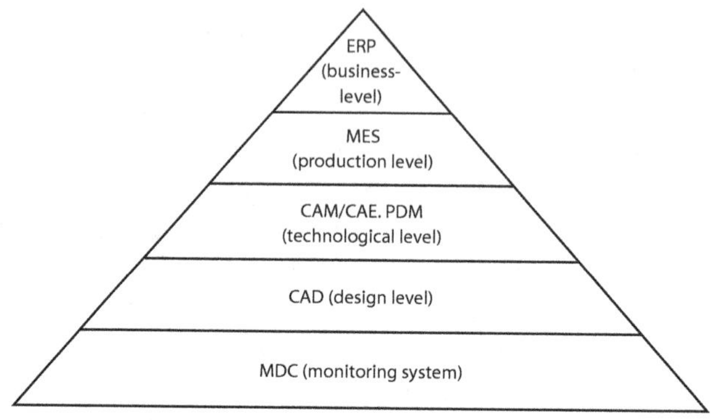

Figure 6.6. Five-level model of digitalization of production

as well as software with data for creating live digital models that are updated and changed (as its characteristics change) aimed at up-to-date representation of the state of a digital twin in real time.

A distinctive feature of this system is the ability to learn on its own, using sensor data that transmits various aspects of its working state, received from workers with relevant knowledge of the high-tech industry, from other similar machines, as well as from larger systems and environments to which the learning system belongs. Besides, digital twins integrate historical data from past machine use into a factor model.

Modern organizations can get a tangible economic benefit from the creation and use of a digital twin for the developed products at the design and development phase. After preparing the design and technological documentation, a digital twin must be created in a way to be the most accurate digital copy of the developed product. The digital twin then goes through all the stages of the life cycle of a physical product. Ideally, a digital twin of a product exactly matches its physical twin and reproduces its state at any time throughout the entire life cycle. Thus, the organization receives a wholesome product lifecycle management technology.

The economic benefit from using a digital twin is determined by the following situations

- a digital twin improves the results of collaboration between designers and technologists. At the moment, there is a common situation in which designers and technologists working in different computer systems face difficulties when working together. Designers transfer their developments to technologists who try to create technological processes in technological systems to which they are accustomed, as a result of which information desynchronizes and error rates may rise. With the use of a digital twin, the error rate due to misunderstandings between designers and technologists is minimized;

- a digital twin of the product allows to optimize the choice of location and manufacturing technology, as well as allocate the necessary resources;
- the use of digital design and modeling allows an organization to avoid situations in which it is necessary to change the product characteristics due to mistakes or obtaining new, previously unaccounted information. Changes in product design entail significant costs that are spread throughout the development phase. Moreover, the later the development stage at which changes are made, the higher the level of expenses that an organization will incur. A digital twin allows an organization to focus the main changes (and associated costs) at the initial stages of product development and thereby significantly minimize total costs, cut costs, and ensure the creation of high-tech products of the future in the shortest possible time;
- testing on a digital twin saves money on the construction of expensive devices used to test the physical properties of a product sample.

The practice of operating PLM systems shows that the main factors in reducing costs in design and production are improving the quality of products, reducing the time for their development and production, manufacturability of design. These factors become sources of additional profits generated according to the scheme shown in Fig. 6.7.

Thus, the effect of additional profit due to PLM is:

$$E_{PLM} = E_{PLM}^{qual} + E_{PLM}^{time} + E_{PLM}^{exp}$$

where E_{PLM} is the effect of additional profit due to PLM systems; E_{PLM}^{qual} is part of the effect obtained from improving the quality

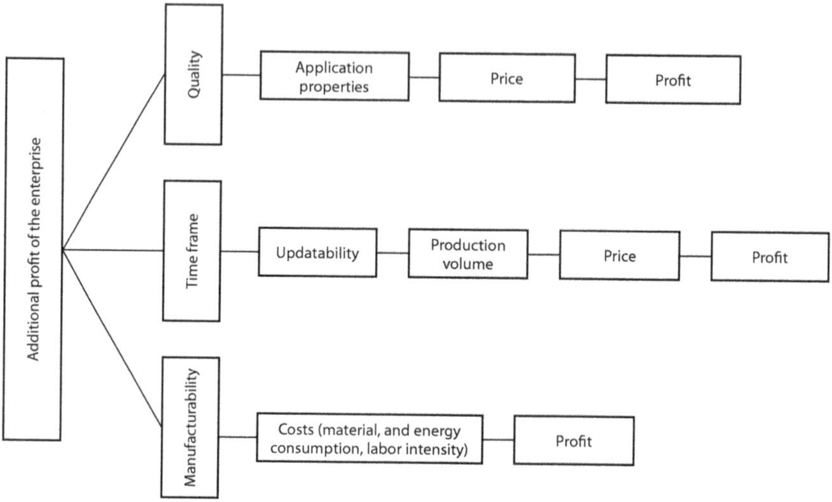

Figure: 6.7. Sources of additional profit of an organization, obtained through the PLM systems

characteristics of products; E_{PLM}^{time} is part of the effect obtained from reducing the production time; E_{PLM}^{exp} is part of the effect obtained from reducing the expenses.

Practice shows that the most significant element of savings from PLM implementation is cost reduction at all stages of the life cycle, covering design, technological preparation of production, testing, control and operation. The life cycle of products manufactured in modern digital organizations is shown in Fig. 6.8.

Factors (sources) of reducing the total costs from the implementation of PLM systems can be divided into two types:

- direct expenses that determine cost reduction at the stage of design;
- indirect expenses that involve cost reduction in the next stages of production and operation.

In this regard, an important issue is to identify the main factors that determine the impact of PLM on the total costs of performing the stages of design and technological preparation of production (direct expenses) and the stages of production, testing and operation (indirect expenses).

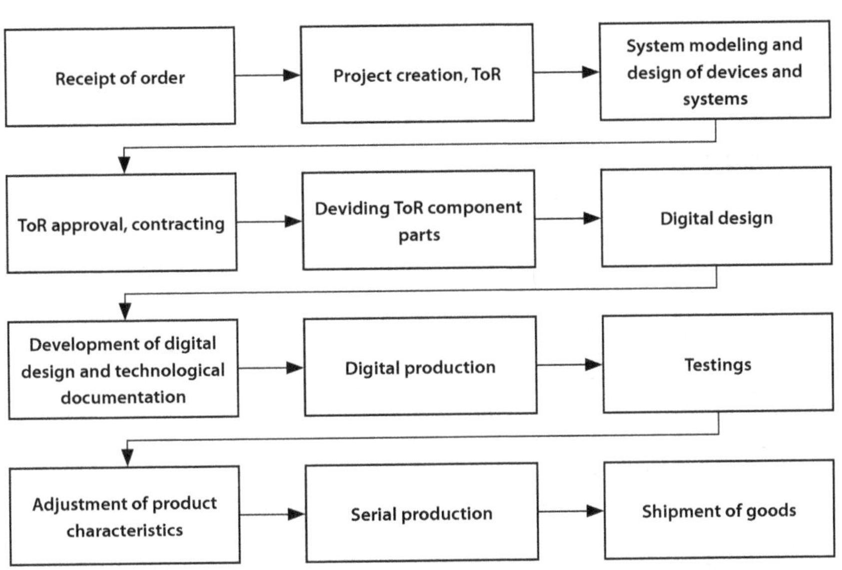

Figure 6.8. Life cycle of modern high-tech products

Analysis shows that the list of determining factors at each stage of the life cycle of a new product is rather limited and can be summarized in table. 6.3.

Assessing the impact of the main factors of production on total costs requires the development of complex deterministic or statistical models. For example, the effect of dimensional design tolerances on cost reduction requires the use of multivariate statistical models. We propose to evaluate the techno-economic efficiency of PLM systems based on the following reporting indicators:

- level of design automation;
- share of industrial products manufactured according to the documentation made by means of computer-aided design in the total volume of production;
- conditional reduction in the number of employees of research institutes (SRI),design bureaus (DB) and the main production as a result of the use of PLM;
- reduction of the unit cost of a product serially produced under the PLM documentation.

Table 6.3. List of determining factors on the life cycles of new products

Design and technological preparation of production	Production (manufacturing)	Testing and control	Operation
1. Reducing the labor intensity of design and technological preparation of pro-duction (TPP) 2. Reducing the time for design and TPP 3. Improving the quality of design (DD) and technical documentation (TD) 4. Improving the manufacturability of the design 5. Simplification and rationalization of the design 6. Improving production specifications 7. Cost reduction for the design of tools and equipment 8. Performing tasks that cannot be solved by traditional methods (optimization and multivariate calculations) 9. Unification and standardization of object design methods 10. Automation of preparation and control of programs with numerical control (CNC) 11. Creation of a unified database of reference information	1. Saving materials, raw materials and energy 2. Reduction of labor intensity and terms 3. Reduction of depreciation charges, saving capital investments 4. Reduction of material costs for finishing the product, shortening the development period 5. Increase in labor productivity during assembly. Lowering the accuracy class of adjoining parts, reducing the range of parts 6. Reduction of expenditure for repair, examination and maintenance of technological equipment 7. Savings from reducing defects due to the fault of the designer and technologist 8. Reducing the expenditure for man-ufacturing tools and technological equipment	1. Reduction of labor intensity and testing time 2. Reduction of the vol-ume of full-scale tests (due to simulation)	1. Reducing the cost of repair and main-tenance of products 2. Increased operation period 3. Improvement of operational and technical characteristics 4. Improvement of techno-economic indicators

One of the integral indicators of PLM efficiency should also be added to these indicators – the annual and integral economic effect, which becomes mandatory when assessing the feasibility of starting the creation of PLM systems, i.e. when opening a new theme and planning the overall economic effect of the introduction of computer technology in an organization.

Thus, the annual savings from the use of PLM are in the areas of design, production of designed objects and the use of production results.

Improvement of the design process from the use of PLM systems is reflected in changes in economic indicators characterizing the activities of a design organization, namely, in the growth of the productivity of designers, reduction in the cost of design, an increase in the volume of design. The improvement of these indicators is provided by the use of specialized software and hardware complexes and high-speed computers, mainly by reducing labor costs and reducing the design time.

PLM makes it possible to improve the quality of design solutions through their optimization, multivariate design, more accurate calculations. Improving the quality of design solutions is manifested in the improvement of the techno-economic indicators of the activities of design, technological organizations, organizations that have mastered the production using PLM, namely in reducing resource costs at facilities designed for one year, in reducing the cost of manufacturing facilities designed using PLM, reducing the operating costs of PLM-designed facilities.

Economic efficiency indicators are intended to substantiate the need and feasibility of creating and using PLM, choosing the best version of the system, its hardware and software, substantiating plans for the development of PLM systems, evaluating the activities of organizations.

The overall effect of using PLM systems is determined by the following provisions:

1. the annual economic effect;
2. the payback period;
3. the estimated rate of economic efficiency.

Let us focus on the main methodological and methodical aspects of assessing the level of automation of design. There are a number of methods for assessing this indicator, two of which are the most complex in terms of their content and the required background information:

1. the level of automation is estimated as the ratio of the number of tasks solved by means of automation tools to the total number of tasks of a typical set in the process of of creating a product;
2. the level of automation is estimated as the ratio of the volume of the automated part to the total volume of a given type of design, expressed in labor intensity.

The first approach to assessing the level of automation is very simple, but extremely pragmatic, since it does not consider the complexity of individual tasks and procedures of design, ways of its implementation (for example, whether a particular task is solved in batch or in a dialog mode, and, moreover, whether it is solved by the developer or a user of the automated system) and many other factors.

This method can be applied under certain conditions to approximate the level of automation. Such conditions should be the development of classifiers of tasks and procedures in the design and, most importantly, the development of a reasonable system of correction (normalizing) coefficients for complexity, methods of performing these procedures etc.

The second method of assessing the level of automation is more advanced, since in this case an objective indicator is used as the background information for calculations – the labor intensity of automated and non-automated design. Labor intensity, in turn, indirectly or directly affects the timing, productivity, the number of working personnel and even the quality of a product, if we bear in mind that for the same time in an automated mode it is possible to view in detail a much larger number of design options than in the traditional way. The following three-level classification is proposed to determine the achieved level of automation of design by division (%):

$$A_{ach} = \sum_{i=1}^{N} A_{ach}\, B_i \cdot 10^{-2},$$

$$A_{iach} = \sum_{j=1}^{M} A_{ijach}\, B_{ij} \cdot 10^{-2},$$

$$A_{ij\,ach} = \sum_{k=1}^{L} A_{ijkach}\, B_{ijk} \cdot 10^{-2},$$

where A_{ach} is the achieved level of automation in the division; A_{iach} is the achieved level of automation in the i-th division; A_{ijach} is the achieved level of automation in the division of the j-th stage of the i–th work; is the achieved level of automation in the k-sub-stage division of the j-th stage of the i–th work; A_{ijkach} is the share of the i-th work in the total amount of work; is the share of the j-th stage of the i–th work; B_i is the share of the the k-sub-stage division of the j-th stage of the i –th work; B_{ij} is the amount of work performed when creating a product; is the number of stages in the i –th work; B_{ijk} is the number of substages at the j-th stage.

The values B_i, B_{ij}, B_{ijk} are determined for all stages of work carried out at the stages of creating products of new technology. It is allowed to enlarge or detail the work performed at the stages of product creation.

The value A_{ijkach} is determined from the ratio:

$$A_{ijkach} = \frac{C_{ijkaut}}{C_{ijk}} \cdot 100\,\%,$$

where C_{ijk} is the full amount of the the k-sub-stage division of the j -th stage of the i –th work, performed without the use of automation tools, person-day; C_{ijkaut} is the part of , for which automation tools are used (in terms of C_{ijk} the volume of the same work performed in a non-automated way), person-day.

Determination of the maximum achievable level of automation in a department is made by the formula (%):

$$A_{max} \sum_{i=1}^{N} A_{maxi} B_i \cdot 10^{-2},$$

where A_{maxi} is the maximum achievable level of automation of the i-th work,

$$A_{maxi} = \sum_{j=1}^{M} A_{maxij} B_{ij} \cdot 10^{-2},$$

A_{maxij} is the maximum achievable level of automation of the j-th stage of the i-th work,

$$A_{maxij} = \sum_{k=1}^{L} A_{maxijk} B_{ijk} \cdot 10^{-2},$$

A_{maxijk} is the maximum achievable level of automation of the k-sub-stage division of the j-th stage of the i-th work.

The indicators B_i, B_{ij}, B_{ijk}, A_{maxijk} are determined by experts, approved by an organization's management and agreed with the leading PLM organizations of the industry and the main subordination departments.

The calculation of the level of automation of design and technological preparation of production at an enterprise is carried out by the formula

$$A = \sum_{z=1}^{\Pi} \frac{A_{achz} N_{at\,z}}{A_{max\,z} N_D} \cdot 100\,\%,$$

where D is the number of design departments in an organization (except for departments whose work is not automated and automation is not provided), units.; $N_{at\,z}$ is the number of engineers and technicians of the z-th design department, person.; N_D is the number of engineers and technicians of the D-th design departments, person.; $A_{ach\,z}$ is the achieved level of automation of z-th department, %; $A_{max\,z}$ is maximum achievable level of automation of z-th department, %; where

$$\frac{A_{ac\,h\,z}}{A_{\max\,z}} \cdot 100\% \leq A_{\max\,z}.$$

Thus, the level of automation of design and technological preparation of production in an organization is calculated as

$$A_{org} = \sum_{g=1}^{M} A_g \delta_g,$$

where δ_g is the share of design performed by g-organizations in the total amount of design work of an organization, %; M is the number of organization; A_g is the level of automation of design and technological preparation of production of a g-organization, %.

Of course, the PLM system helps to quickly achieve the main goal of modern designers – to create products with high competitive advantages. Among other things, PLM brings attention to an issue that is often forgotten in the process of designing a product, namely, optimizing the production cost, and eventually its competitive price.

Consider other tools that can improve the efficiency of developing a digital product twin. A highly adequate "smart model" given the specifics of a particular product is called a digital twin of production.

If we simulate the behavior of a product at the next stages of the life cycle – operation and disposal, then another tool appears that increases the efficiency of the digital twin of a product. This tool is called digital product shadow and allows simulating situations such as repairs, accidents, scheduled maintenance, upgrades, etc.

The digital twin plays a key role to play in Industry 4.0, and the world's PLM development leaders have been developing software tools for the decade to enable the process of creating a digital twin. PLM developers offer different approaches to the definition of a digital twin. For example, Siemens PLM Software sees it as the intersection of four areas – product development, production planning, manufacturing facilities and the real world, with a particular focus on manufacturing. Dassault Systèmes prefers to use the term "virtual twin", which is a development of a system engineering strategy. A virtual twin allows

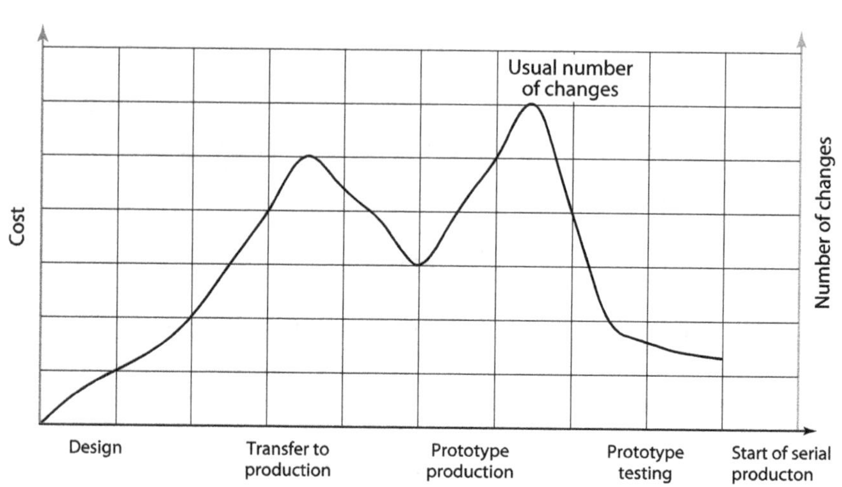

Figure 6.9. Dynamics of development and production cost
with traditional approach

the development team to create a product that combines mechanical, electrical, electronic, hydraulic and other systems, and then test and handle it by studying its behavior under various conditions (loads, vibrations, software operation, control systems, etc.)[23].

Considering the issue of optimizing an organization's costs for the development and manufacturing of new products, it should be noted that the cost of production is directly dependent on the quantity and amount of expenditure for its creation. In this regard, we investigate the dynamics of production costs with the traditional approach (Fig. 6.9) and with the digital approach (Fig. 6.10).

When comparing the two graphs (Fig. 6.9 and 6.10), it becomes obvious that with the traditional approach, an organization bears the most significant expenses related to the transfer of design documentation to production (due to misunderstandings that arise between designers and technologists, which was mentioned earlier) and at the stage of production of a prototype (due to obtaining new information). With a digital approach to production, most of the

---

23 Stackpole B. Digital Twins Land a Role In Product Design. – URL: http://www.deskeng.com/de/digital-twins-land-a-role-in-product-design/

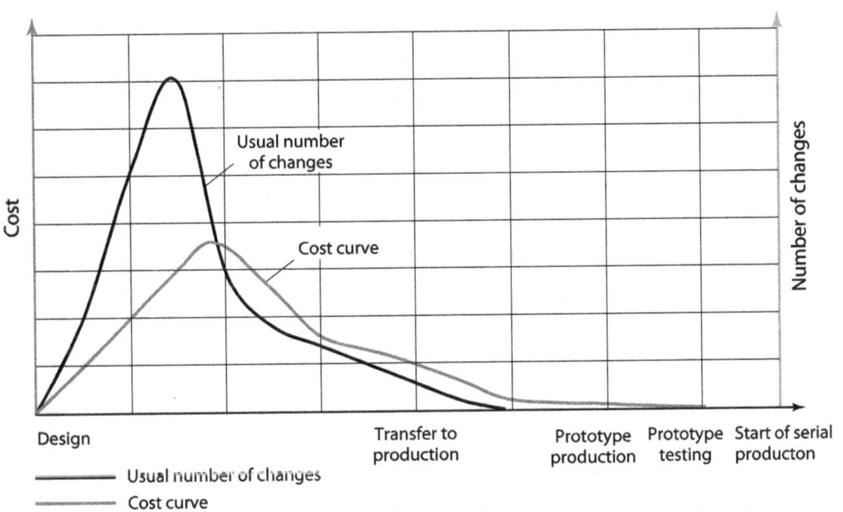

Figure 6.10. Dynamics of development and production cost
with digital approach

expenses come from the product development stages. This feature is justified by the use of a digital twin of a product, which allows to make the necessary changes and adjustments before the direct production of a physical sample of the product.

The amount of production costs in the traditional approach is significantly higher than the costs in the digital approach. It is worth considering the difference in time required to bring products to market: with the digital approach, significant time is required for preparing tests, since they are carried out virtually on a digital twin, while with the traditional approach, all tests are carried out on a physical sample of the product and, accordingly it takes time not only to prepare and conduct the test itself, but also to produce a prototype. Due to the use of a digital twin, the full-scale test period can be shortened by conducting virtual tests on a digital twin of the product.

It also follows that the digital approach reduces the cost of production and testing of a prototype product, because when conducting virtual tests on a digital twin of a product, it is possible to improve many techno-economic characteristics of the product and improve it without expenditure for manufacturing a physical sample of the product.

Thus, the need to use PLM-systems and a digital twin in the activities of modern knowledge-based organizations is dictated by the significant positive economic effect of the use of such systems.

6.3. METHOD OF COMPREHENSIVE ASSESSMENT OF THE EFFECTIVENESS OF THE PRODUCT UPDATE PROCESS FOR THE SPECIFIED TECHNICAL-ECONOMIC PARAMETERS

In modern conditions, when many manufacturers with a high level of competence operate in the markets of high-tech products, the issue of effective product update becomes especially important for achieving the competitiveness of a high-tech organization as a whole. In order to assess the effectiveness of product updates, managers are forced to constantly evaluate proposed changes, in particular, their impact on products.

Often, managers tend to make decisions about introducing new functionality (and other changes) into the product, without fully assessing the need and feasibility of these adjustments. Inappropriate management decisions lead to the creation of an uncompetitive product with zero profit or loss.

The Samsung T9000 AVK refrigerator is one example of a product for which the addition of high-tech capabilities has become disastrous. The model was presented by Samsung in 2013 at the largest exhibition of the latest high-tech products of the Consumer Technology Association and immediately attracted attention with an unusual four-door design and a touch screen built into the refrigerator panel. However, in practice, this high-tech screen, which significantly increased the market value of the refrigerator, did not bring any benefit to consumers. For this reason, this model of the refrigerator did not gain popularity, while the RF9000 model with the same design but not equipped with a touch screen, is successfully sold today. The T9000 FDR model, presented on the American version of the Samsung website, is not available for purchase in any store.

Figure: 6.11. Unsuccessful project of the RAH-66 "Comanche"
family helicopter model[24]

Of course, the development and production of consumer products (for example, refrigerators, televisions, etc.) equipped with high-tech functions requires less costs from a manufacturer than the production of high-tech products (in particular, helicopter equipment, aerospace technology, etc.). For example, the program for the development and construction of multipurpose reconnaissance and attack helicopters RAH-66 Comanche, launched by the United States in the 1990s, cost the country $ 2.8 billion. At the end of the 20th century, this program became the most ambitious of all the adopted projects of the US Army: the models of the RAH-66 Comanche family (Fig. 6.11) were to replace the AN-1 Cobra, Bell UH-1 Iroquois, Bell OH-58 Kiowa models and the OH-6 Caius, which made their maiden flight in the middle of the 20th century.

As part of the program, it was planned to build 4-6 thousand helicopters of the RAH-66 "Comanche" family, equipped with the latest technology, for an amount of 24-36 billion dollars. However, since the launch of the program, the requirements have gradually decreased: by 2000, it was planned to build only 1292 helicopters of the RAH-66 "Comanche" family, and by 2010 the project was

---

24 Source: http://politrussia.com/vooruzhennye-sily/top-5-samykh-225/.

closed due to the war in Iraq, which required significant spending from the US budget. Thus, the program, which resulted in the production of a single prototype of the RAH-66 "Comanche" helicopter family, cost the United States $ 9 billion and never reached the goals set in the 1990s.

Usually, to make an effective management decision about the need for certain changes, it is necessary to answer the following questions:

- how will the technical-economic characteristics of a high-tech product change??
- will the change solve the existing market demand in a more efficient way (at a lower cost or in a shorter time frame) or meet a new market demand that competitors have not yet been able to meet?
- how will the competitiveness of the product change as a result of these changes?
- is it reasonable to invest money in the practical implementation of these changes?

Of course, the answers to the questions posed require thoughtful and thorough reflection from the manager. However, the modern market, characterized by rapid changes, dictates conditions under which the development of long-term and cumbersome business plans becomes irrelevant, while the importance of strategic vision increases significantly.

A manager, when deciding whether to upgrade a high-tech product or manufacture a new one, must take into account the long period of R&D necessary to develop a new product or improve an existing one.

The following method is proposed for assessing the effectiveness of changes in high-tech products.

1. Determine the expected impact of changes.

Different changes made to different types of products will naturally lead to different consequences, for example, to:

- increased demand for the product;
- attracting new clients;
- increasing the satisfaction of existing customers;
- increasing market share;
- coverage of changes made to the product in the media or analytical blogs.

2. Set goals for change.

After determining the expected impact of changes made to a high-tech product, it is necessary to establish the final and intermediate goals of these changes. These goals should not only have a precise formulation that does not allow ambiguity and double interpretation, but also objective assessment criteria to determine whether the goal has been achieved or not.

3. Identify a measurement method to determine whether the change you are making allows you to achieve the goals set in the previous step.

Based on the first step, the expected impact of changes is clear, but it is also necessary to establish a number of criteria by which a manager will be able to measure the impact of changes. For example, if you are considering adding a new feature to a high-tech product and the organization's goal is to have 30% of customers use this feature after purchasing an updated product, you need to establish a feedback channel through which the organization can get information on whether this feature of a high-tech product will be in demand from 30% of customers.

This point seems obvious, however, as practice shows, criteria for measuring the impact of changes is not elaborated out until the direct implementation of the changes. In this case, an organization is able to assess only the effectiveness of the already implemented practice changes: that is, if the change has already been implemented and the goal is not reached, then time and material resources were wasted. A preliminary assessment of the attainability of the established goals by implementing the assessed change allows to receive guarantees that the set goals will be achieved or to detect the inconsistency of the implemented change with the goals pursued by a high-tech organization. In the second case, the organization has the option to select a different change to achieve the stated goals.

4. Measure the impact of the change and objectively assess its effectiveness based on the results of implementation.

After implementing the changes, it is necessary to compare the actual results obtained with the expected ones predicted at the previous stage. The following questions have to be answered:

- are the results set in the third stage of the method achieved?
- if the goals were not achieved, what was the reason for this?
- are there any further changes that need to be made to the product?
- what experiences should be noted as positive?

If possible, this experience is recommended to be used consistently with future product changes.

Every product is created with a specific purpose; therefore, when determining the need for a particular change, a manager must correlate the proposed changes with the purpose of the product.

In the event that a high-tech organization wants to be competitive, managers' main goal should be targeted product update, and not just making changes for the sake of changes. Managers need to bear in mind that the goal of an organization is not to change the product itself, but to increase the organization's revenue. For this reason, instead of spending money and resources on implementing 10 mediocre changes, they should focus on implementing one change, the benefits of which for consumers will exceed the total utility of 10 functions.

Speaking about measuring the impact of changes and objectively assessing their effectiveness, it should be noted that if a particular goal was not achieved does not mean that the change was futile. For example, if, as a result of introducing a new function into a high-tech product, an organization's income increased by 5% instead of the planned 10%, this is still a positive trend. In such a situation, it is necessary to objectively assess whether the goal was realistic and what needed to be changed in order to achieve it. The conclusions drawn can be used by the manager in the future.

Note that some changes are much more difficult to measure than others. Besides, it is impractical to measure every small change in the product. However, without using this method for significant changes, managers actually act intuitively, blindly, since they are not guided by objective indicators. Setting and tracking a number of metrics for achieving the goals of updating a product or changing it will enable creating a better product and identify possible problems before they occur.

The fundamental point in applying this method is the consistency of the goals and expected consequences in the implementation of the change. The alignment process may be difficult and it is likely that agreement on all aspects will not be possible, but it is advisable to identify the inconsistencies in the objectives during the alignment process. For example, a marketing manager may feel a need for product upgrades that attract new customers, while the chief engineer wants to make performance-enhancing changes that will retain existing customers.

However, the application of the proposed method (namely, paragraph 1 and paragraph 2) will allow for greater coherence and understanding between an organization's departments, since the discussion process usually focuses not on the details of implementing these changes, but on their necessity.

The habit of applying this method, even when it is impossible to fully track or measure changes, will help form managers' mindset that satisfies the needs in shaping products of the future, in achieving advanced development, in which an organization can become the most competitive in the existing market or create a new market and be a monopolist for a while.

CHAPTER 7. GLOBAL COMPETITIVE EXCELLENCE OF AN ORGANIZATION BASED ON THE CREATION OF PROMISING HIGH-TECH PRODUCTS IN THE CONDITIONS OF DIGITAL TRANSFORMATION

7.1. ASSESSMENT OF THE COMPETITIVENESS OF THE MANAGEMENT SYSTEM FOR LAUNCHING NEW PRODUCTS IN THE DIGITAL ECONOMY

The effectiveness of the management system for new products launch is largely determined by the effectiveness of the organization of the project management process for creating a product that dominates the market at all stages of its implementation. In this case, the competitiveness of the system can be seen in results of monitoring changes in the competitiveness of the product following the implementation of measures to preserve it or increase competitive advantages at the stages of the life cycle.

We consider a system as competitive if it allows you to maintain and increase competitiveness of a product created as part of a project at the stage of assessing the techno-economic image, design, preparation of production, production and the moment the finished product is brought to the market, and the competitiveness of an organization as a whole, since the management system of creation and launching of new products is part of an integrated organizational and economic system for managing the economic activities of organizations.

Assessing the competitiveness of the system for managing the production consists not only of the characteristics of the system itself, but also of the analysis of the competitiveness of products manufactured under the control of the analyzed system.

To analyze the competitiveness of the management system for launching new products, the method of analyzing the competitiveness of objects is used. It must be carried out using a single method, allowing to obtain numerical characteristics of competitiveness for individual objects. Accordingly, when analyzing the competitiveness of complex objects with a certain hierarchy, such as high-tech corporations, methods for analyzing the competitiveness of hierarchical structures will be applied. These methods allow us to build a quantitative assessment of the competitiveness of a complex object.

Let us consider the analysis of the competitiveness of a set of individual objects, which are the products of a high-tech corporation. When using this method the dynamics of changes in the competitiveness of products of a high-tech corporation is studied in relation to the time that determines the start of work on the production of new products.

Thus, the proposed technique allows us to assess the change in the vector of quantitative indicators of product competitiveness, as well as an organization as a whole. Based on this vector, it can be concluded that the measures for setting up new products in production are effective due to the use of the decision support system (a management system for the production of new products). The developed technique is based on computational procedures, and not on expert assessment methods, which ensures its objectivity. On the other hand, the calculation formulas include a sufficient number of parameters, which allows you to configure the computational procedures in an optimal way.

The methodology for assessing the competitiveness of management decision support systems when launching new products is aimed at developing unified computational procedures for obtaining objective information about changes in the competitiveness of products and organizations as a result of the use of these management systems.

This technique allows us to solve the problem of automating the measurement of the effect of using control systems for setting up new products, since it is based on computational procedures that can be effectively implemented in the form of an information and computer program complex.

The result of calculations based on the method of assessing the competitiveness of an organization due to the systems launching new products will be the calculated vector of the integrated assessment of the organization's competitiveness. Based on the obtained indicators, we can conclude that the ability of a high-tech corporation to effectively manage the introduction of new products and thereby manage its own competitiveness.

For effective implementation of this method, it is necessary to have data on the competitiveness of products, the level of development of adaptive and flexible production systems, the level of competence of an organization, the level of competence of related parties, the level of interaction with customers in terms of forming a technical task for product development and determining consumer preferences.

The first stage of quantitative assessment of an organization's competitiveness through systems for launching new products is the procedure for analyzing the competitiveness of planned products based on the integrated competitiveness indicator. This model can be used to determine the competitiveness of both finished products and components created for their use in the production of more complex products.

As the initial data of the algorithm for determining the integral indicator of product competitiveness, we will consider the following parameters:

S_0 – the cost of financial investments made by the organization to create products and start production for its release;

S_1 – unit cost;

S_F – the price of similar products on the market (if there are no analogues, then $S_F = S_1$);

N – the forecasted number of units of products to be sold on the market;

T – the time required for development, the organization of production and start of sales;

T_0 – the planned time of operation of the product;

α – a coefficient from 0 to 1, reflecting the importance of the waiting time for output;

R – internal risks arising in the process of production by the company;

F – external risks accompanying the sale of products on the market.

After forming a list of parameters, it is necessary to calculate their numerical values. In this case, you can use natural values or scales.

Let us focus in more detail on the calculation of the product innovation index Q. Following the method of solving this problem, we will assume that this indicator is a function of particular indicators of innovation, which are understood as key technical characteristics of the product. So,

$$Q = f(q_1, q_2, ..., q_n),$$

where q_i, $i=1,...,n$ is a particular indicator of product innovation, representing some key technical characteristic. A particular competitiveness indicator q_i is a value that takes values from the segment [0; 1]. Thus, its value can be found by the formula

$$q_i = \frac{u_i}{U_i},$$

Where u_i is the technical characteristic of the product; u_i is the same technical characteristic of the market leader or the maximum possible value of the characteristic in accordance with the current level of scientific and technical progress.

The indicator of product innovation Q can be found based on particular indicators of product innovation taking into account the weight coefficients of the importance of indicators of innovation:

$$Q = \sum_{i=1}^{n} \beta_i q_i,$$

where β_i – are the weight coefficients of particular indicators of product innovation, and

$$\sum_{i=1}^{n} \beta_i = 1.$$

"External risks" are estimated by a dimensionless value from the segment [-1;1]. The value -1 corresponds to the maximum negative impact of external (political, economic) factors on sales in the target market; the value of 0 corresponds to the absence of external factors of friendliness; the value of +1 corresponds to the most positive influence of external factors on sales on the market. The remaining values are intermediate characteristics of the influence of external factors on potential sales of new products on the market.

"Internal risks" are equal to the probability from 0 to 1 output failure.

The indicated parameters must be set for each "product – target market" pair. After that, an objective indicator of product competitiveness is calculated for each product on the market using the following formula:

$$IQ = \left(\frac{Q \cdot S_F}{\frac{S_0}{N} + S_1} - \alpha \frac{T_0}{T} \right) \cdot (1 - R) \cdot (1 + F).$$

This formula expresses the numerical competitiveness index of the products in question in the market. The integral indicator of IQ competitiveness can take values either greater than 1 or less than 1. For $IQ > 1$, we can say that the competitiveness of products created on the market is higher than the level existing on the market.

The competitiveness of an organization is directly related to the competitiveness of its products. At the same time, the competitiveness of an organization depends on the entire range of products.

If integral indicators of the competitiveness of all types of products $IQ_1, IQ_2, ..., IQ_N$, are known where N is the number of types of products, then the integral indicator of the competitiveness of the entire set of products produced by an organization can be found by the following formula:

$$IQ = \sum_{i=1}^{N} \omega_i IQ_i ,$$

where ω_i is the weight coefficient characterizing the contribution of each type of product to the formation of an integral indicator of an organization's competitiveness. As the weight coefficient corresponding to a certain type of product, we will consider the share of revenue from sales on the market in the total revenue of an organization:

$$\omega_i = \frac{V_i}{V},$$

where V_i – revenue from sales of products of the type i; – the total revenue of the organization. Thus, .

$$\sum_{i=1}^{N} \omega_i = 1 .$$

The competitiveness of an organization is determined not only by the competitiveness of its products, but also by a number of other factors that are determined by the capabilities of the system for launching new products:

- the level of development of the adaptive production system (K_1);
- the level of competence of the organization (K_2);
- the level of competence of related parties (K_3);
- the level of interaction with customers in terms of creating a technical task for product development and determining consumer preferences (K_4).

To assess the listed factors of competitiveness, you can use well-known techniques and algorithms. The calculation result by the method must be normalized to unity in order to obtain dimensionless quantities.

With known values of the competitiveness index of the entire set of products IQ and estimates of other factors K_1, K_2, K_3, K_4, the integral indicator of the organization's competitiveness will be written as:

$$IQO = w_0 \cdot IQ + w_1 K_1 + w_2 K_2 + w_3 K_3 + w_4 K_4,$$

where w_i are the weight coefficients of the organization's competitiveness factors. Thus,

$$\sum_{i=0}^{4} w_i = 1.$$

We will consider an organization that uses a decision support system with an integral indicator of its competitiveness $IQO \geq 1$ for the production of new products as competitive.

The proposed algorithm for an objective assessment of an organization's competitiveness due to systems for launching new products allows you to quickly get a dimensionless index of its competitiveness.

The successful implementation of a project to create a highly competitive product that could displace existing products or create a new market directly depends on the availability of key competencies that can be used to obtain unique technical, economic and consumer characteristics of new products. Thus, the foundation of advanced development is the competitiveness already formed by an organization. In a conceptual aspect, it is a system of knowledge, skills, methods and tools for managing an organization's activities, which allows achieving strategic goals by implementing breakthrough projects to create highly competitive products. Organizational project management can be applied both at the organization level and when implementing a specific project.

7.2. FORMATION OF A CONTROL SYSTEM FOR LAUNCHING NEW PRODUCTS IN THE CONTEXT OF DIGITAL TRANSFORMATION

Currently, the adoption of effective managerial decisions at each stage of the implementation of an innovative project is becoming one of the foundations for creating promising products, forming their competitive advantages, ensuring their dominant position in the market, and creating value for an organization. The effectiveness of the process of setting up new competitive products is the key to improving the economic stability of an organization, its competitiveness, provided by the competitiveness of products, compliance with world-class standards, which is especially important for high-tech companies.

At the same time, the competitiveness of products varies at all stages of their life cycle depending on factors of the internal and external environment, and therefore there is an urgent and important task to manage the processes of development and production that will be competitive by the time the products enter the market. The solution to this problem in modern conditions should be based on effective management decisions.

Over the past few decades, management systems for launching new products have been significantly developed. They have evolved from automated systems used to simplify calculations, to complex organizational and economic mechanisms combining hardware and software, as well as artificial intelligence to process large volumes of data, study and analyze results, identify new opportunities and effectively implement strategies.

The new product launch management system is an important tool that can quickly assess a number of parameters, conduct an in-depth analysis, identify future trends, and suggest possible actions. This means that it is able to visualize future trends in a particular area, offering useful business ideas that create certain competitive advantages for the organization. This control system for new products launch is constantly developing. New technologies, new approaches and changing needs of the business determine this continuous dynamic development. In the course of this process, the competitiveness of the management system for launching new products is formed, which means the ability of the system to recommend solutions that will achieve the greatest increase in the competitiveness of products, optimize resources for its production at all stages of the life cycle, reduce the time of its production, analyze risks based on their monitoring by weak signals, determine the degree of their influence and compensate for the consequences by implementing the adopted list of measures.

Moreover, the mentioned control system should have the ability to adapt flexibly depending on changes in production conditions, the development of science and technology, the emergence of new technologies and other factors of the internal and external environment.

In modern conditions of development of the digital economy and creation of new control systems using artificial intelligence technologies, control systems for setting up new products are also being transformed. Studies have shown that, in general, they can be divided into several classes depending on the direction of the tasks they solve.

Problems arising in the management of complex technical and technological systems, such as production systems in the high-tech industry, can be divided into the following classes:

- monitoring or assessment of the situation;
- forecast of system behavior in the usual (regular) mode;
- forecast of emergency situations;
- development of possible solutions;
- the choice of the most satisfactory solution.

At the same time, it was found that existing systems for managing the launch of new products are usually focused on solving specific tasks in accordance with the costumer's request, which makes it necessary to adapt them or develop new ones to solve certain tasks.

Besides, a study of the systems used in the practical activities of organizations showed that they do not allow assessment and ranking of innovative projects at the pre-investment stage; technical, technological and other potential; competencies of an organization in terms of its ability to create innovative competitive products on the market, and are also not aimed at solving the problems of managing the competitiveness of a product.

The study of current scientific literature regarding management systems for setting up new products with different functionalities shows that the competitiveness of systems is usually related with questions of the effectiveness of the implementation of management decisions made with the support of these systems, which leads to the creation of competitive products or services demanded on the market and holds a high market share.

It follows that the effectiveness of the management system for launching new products is calculated based on an assessment of the competitiveness and market demand for the product manufactured as part of the project managed by this system.

This system should include subsystems for managing the processes of creating promising products that can provide a leading position in the market, as well as analytical tools based on modern technologies of the digital economy, including methods for processing and analyzing big data. Based on these tools, competitive advantages are created for the control systems for introducing new products, as well as assessing the status of promising markets, segments of product sales, changing consumer preferences, assessing the competitiveness of products sold in the markets in the current and forecasted situation, evaluation of the competitive value of the created products and formation of competitive advantages of products that will ensure their dominance in the market.

This is why our approach to determining competitiveness is related to solving three problems of assessing competitiveness:

1. creating an image of a new product launch management system that could provide recommendations for implementing control actions on a project or product at all stages of its life cycle;

2. development of the methodological apparatus of the system using economic and mathematical modeling, on the basis of which effective management decisions can be made;

3. assessment of technical, economic and functional characteristics of the system and their comparison with similar characteristics of existing systems.

Thus, the competitiveness of production control systems for setting up new products is qualified by a combination of technical and functional data of the information system and indicators of the product creation management system at all stages of its life cycle. When forming the list of indicators characterizing the specified management system as informational, its main indicators will be directly technical and functional parameters (Fig. 7.1).

The functional characteristics (capabilities) of a system that can provide it with competitive advantages determine the formation of new needs on the part of users of such systems:

- substantiation of managerial decisions on financing projects at the pre-investment stage;
- implementation of the process of effective management of the development and production processes of products that can occupy a dominant position in the market;

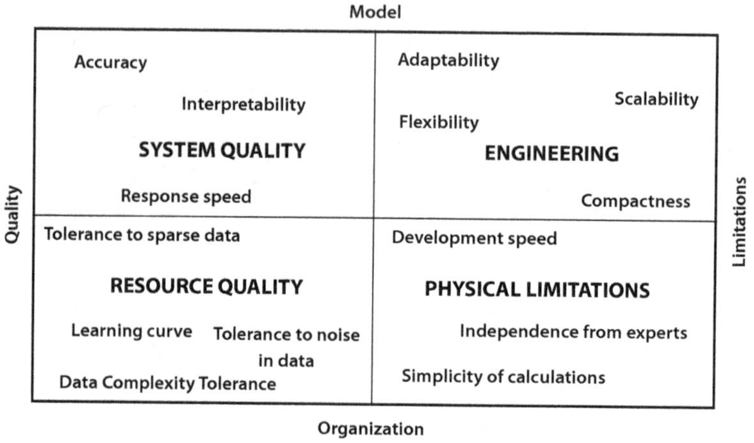

Model

Accuracy **Interpretability** **SYSTEM QUALITY** Response speed	Adaptability Scalability Flexibility **ENGINEERING** Compactness
Tolerance to sparse data **RESOURCE QUALITY** Learning curve Tolerance to noise in data Data Complexity Tolerance	Development speed **PHYSICAL LIMITATIONS** Independence from experts Simplicity of calculations

Quality — Limitations

Organization

Fig. 7.1. Technical and functional parameters

- integration with other information systems, including PLM;
- the use of OR methods (Operational Research – business modeling) integrated into decision support systems in order to provide a fairly realistic picture of certain processes for a decision maker.

The new product launch management system implements the management process in a structured form and helps decision-makers determine an acceptable solution for a specific task, while forming a stable economic position of the organization in the market and increasing its competitiveness. At the same time, competitive advantages are created by:

- shorter decision-making period: the time spent studying data and comparing possible courses of action is significantly reduced. The decision-making cycle is getting shorter, which allows you to speed up the production process, ultimately reducing the time it takes to bring it to market;
- increasing data objectivity: the human factor in the decision-making process is minimized;

- improving the quality of strategic management: a decision support system is changing the way organizations work. An important concept that uncovers the role of computerized decision making is "value chain management". The decision support system takes into account economic factors, past and current trends to determine costs and profits, as well as total cost;
- flexible response to market changes: adapting to changing market conditions and needs helps to stay ahead of the competition. In fact, this is what makes organizations more flexible and innovatively active, able to respond quickly to changes in market conditions.
- the decision support system, using available information, presents projected revenue indicators and expected market changes in the future;
- reduce the cost of decision-making. Using a decision support system significantly reduces the cost of collecting, sorting, processing, and analyzing data. The cost of storing information, hardware, computer and Internet technologies is significantly falling. This means that the cost of extending decision-making technology even to lower levels of the hierarchy is reduced.

The input data of the system are the characteristics of the techno-economic image of products, projects that require a decision on the possibility and feasibility of their implementation.

The output data of the system are the recommendations for managing product development projects to ensure their high competitiveness and demand in the market, taking into account dynamically changing external and internal factors.

The system's work is based on the use of methods for collecting, processing and analyzing big data, artificial intelligence, economic and mathematical modeling and is aimed at phased assessment of the possibilities and feasibility of further implementation of the product development project.

Automated decision support in the broad sense means at least one of the following functions:

1. providing reference information without automatically generating database queries;
2. providing reference information with automatic generation of queries to databases and operating under conditions of the problem to be solved;
3. graphical visualization of the received reference information and information about decision-making methods;
4. providing recommendations for decision-making;
5. narrowing the search space for a solution by the user;
6. selection and recommendations of the most appropriate solutions, taking given the ranks;
7. modeling the consequences of decision-making.

Thus, the main task that the system should solve is to support the process of creating competitive products that ensure the competitiveness and sustainable development of an organization.

Product competitiveness is a dynamically changing parameter throughout the entire life cycle, and therefore it is necessary to make effective management decisions on its maintenance, starting with the formation of its techno-economic image.

Let us now consider the principles of implementation of the main functional blocks of the system for setting new products, aimed at determining recommendations for the implementation of a project to create competitive products, on the basis of which effective management decisions can be made and implemented.

Subsystem for evaluating opportunities and potential sales volumes of a product on the market. Product ranking by sales volume indicator.

This subsystem is used to analyze whether the product characteristics meets the requirements and wishes of the consumer and identifies the most important technical characteristics of the product that meet the expectations of the consumer and ensures its competitiveness in the market. The initial data for evaluating projects at the pre-investment stage are the results of marketing research that determine consumer expectations, the importance of certain properties, compliance with requirements and expectations, as well as proposals and developments of competing companies. Within this subsystem, it is necessary to have methodological tools that can be used to identify the future needs of organizations and society based on their growing intellectual potential and developing competencies. Such a forecast should be based on the intellectual analysis of big data of the global information space, taking into account the development of the current technological structure and a number of different macroeconomic and industry factors that affect sales. The basis of the subsystem can be the methodological foundations of its creation, which are aimed at solving the problem of determining prospective needs in various types of products in an automated mode. The system allows you to set quantitative indicators of potential deliveries of a specific type of product in the planning period, broken down by the internal and external market, as well as some segments of consumption.

Subsystem for determining the competitive price of the selected product, taking into account uncertainties.

This subsystem as part of the decision support system is necessary to determine and justify the competitive price of the future product, which will be produced in accordance with the parameters of the techno-economic image uploaded to the system.

The subsystem is based on a unified methodological approach to calculating and forecasting the competitive price of a future product.

This subsystem is divided into 3 main blocks:

- determining the competitive price of a product;
- predicting the competitive price of a product;
- determining the cost of a product depending on the price of materials.

The purpose of the subsystem is to calculate the competitive price and its forecasting to determine the boundaries of the cost of the future product so that it can occupy a high market share, as well as setting the price of a multicomponent product based on the analysis and forecasting of the market conditions for materials, industrial raw materials and chemicals used in high-tech manufacturing industry, taking into account variations in external factors in the short and long term. The tasks of the subsystem are:

- selection of the most significant characteristics of the techno-economic image of a product that affect its price;
- defining the boundaries of competitive pricing for a future product;
- issuing recommendations for creating a competitive price for a product;
- analysis of the state of the markets for materials, industrial raw materials and chemicals used in the production of high-tech products;
- building a forecast of prices for materials, industrial raw materials and various chemicals that are used in the production of the product in question;
- decomposition of a multi-component product into separate components and connections between them;
- determining and updating the product price taking into account external factors.

As a result, the subsystem generates the recommended price of the future product, which will ensure its price competitiveness.

The subsystem, solving the problem of determining the above price, uses the following indicators:

- the total price of a multi-component product;
- forecast values of prices for industrial raw materials, materials and chemicals used in the production of high-tech products.

In addition, the subsystem calculates the cost of various sub-components of a multicomponent product.

The subsystem for assessing the competitiveness of the techno-economic image and forecasting the competitiveness of the future product.

The subsystem for assessing competitiveness is based on the methodology for quantitatively assessing the competitiveness of products and monitoring the dynamics of their changes.

Based on the analysis of changes in competitiveness indicators, the management decision support system forms a list of recommendations for:

- priority activities needed to maintain competitiveness;
- activities aimed at creating new competitive advantages;
- adjusting operations that are implemented during the product creation process.

The product competitiveness assessment tool provides an assessment of indicators, including at the early stages of the formation of the techno-economic image of the product, which can be achieved on the basis of a wide range of modern information technologies, knowledge bases and informal methods of synthesis and analysis.

The structure of the proposed mechanism for the preventive assessment of competitiveness (MPAC) includes fundamental, theoretical, methodological and informational components.

The goal of MPAC is to achieve new progressive qualities of the product, ensuring a high level of its competitiveness through the use of knowledge. The effectiveness of innovations is assessed by the market when buying and selling. Under design conditions, when market information is not available, the progressiveness of new technology is based on macroanalytical analysis, the complex elements of which are competitive advantages.

Thus, the fundamental basis of MPAC is the development of conceptual scientific and technical proposals in the form of a set of competitive advantages that reflect achievements in the field of scientific and technical progress and labor productivity. In creating competitive advantages, an object-oriented R&D program and scientific and technical developments in related fields are used, including research work on the system analysis of patent information and the development of technical proposals to improve the technical level of new generation products.

The theoretical and methodological components of MPAC form a scientific and methodological approach to product design based on a preventive assessment of its competitiveness. which distinctive features are the use of an automated interactive design system that allows you to search for the best product samples that meet the requirements for both techno-economic characteristics, and the use of a comprehensive methodology for ensuring competitiveness, including analytical and informal algorithms for analyzing and synthesizing products, adapted to the variant analysis of competitive advantages and practical evaluation of the final results of innovations at the early stages of the formation of the techno-economic image of products.

Subsystem for evaluating the sufficiency of resources to create competitive products.

The problem of resource support is the difficulty of accurately planning the volume and type of this support in the development and production of products that should take a dominant position in the market. It is necessary to have sufficient resources to manage this project. Given that the resource support is multidirectional (for the development and production of high-tech products, financial, intellectual, information, human and other resources are required), it is important to determine the necessary resource support in the complex.

This subsystem should answer the question of the adequacy of resources in the context of creating a highly competitive product that can occupy a dominant position in the market.

The subsystem takes into account the specifics of planning both tangible and intangible resources, which are of equal importance at certain stages of the life cycle of high-tech products.

For example, when implementing a production program for the production of mastered products, it is more important to plan material resources efficiently, and at the design stage and in pilot production, more attention should be paid to the effective use of developers' competencies.

Resource planning is implemented in three stages. At the first stage, the necessary resource support is assessed according to the classical scheme. At this stage, sufficient amounts of financial, material, labor, time, information and other resources are described both at the stages of research and development of high-tech products, as well as at the stages of pre-production and direct production. At the second stage, the subsystem includes the factors of innovation that take into account the specifics of high-tech production. At the third stage, the forecast of the actual required resources is based on the adaptive model.

This subsystem allows you to plan resources based on the observed trends in the required resource provision over the course of the year or other arbitrary time frames.

The following principle should be laid down to create the appearance of a management system for setting up new products: the system at different stages of the life cycle should influence the project or product as an information system, the data of which are used by the head to make managerial decisions that ensure the creation and maintenance of techno-economic characteristics project or products incorporated as source materials. For this, the system should have multifunctional capabilities that allow assessing the state of the project at different stages of its life cycle, and under certain conditions, offer a description of the process of improving the parameters of the project or product in order to maintain the given competitiveness or create new competitive advantages.

7.3. Fundamentals for creating global competitive advantages for an organization

The objective of any organization operating in a market environment is stable economic development, for which the appropriate business strategies are formed to find a way to achieve competitive advantages over competitors' products and services in the market.

Competitive advantage refers to the priority that an organization receives (in comparison with its competitors) which allows it to offer consumers products with better techno-economic characteristics, at a lower price, or by providing greater benefits and accompanying services that justify the high price of the product.

M. Porter considered four business strategies that allow organizations to achieve competitive advantages (Fig. 7.2).

At the same time, the strategy of differentiation and the strategy of cost leadership imply obtaining competitive advantages in a wide range of market segments, while the strategies of focusing on costs and differentiation mean obtaining competitive advantages

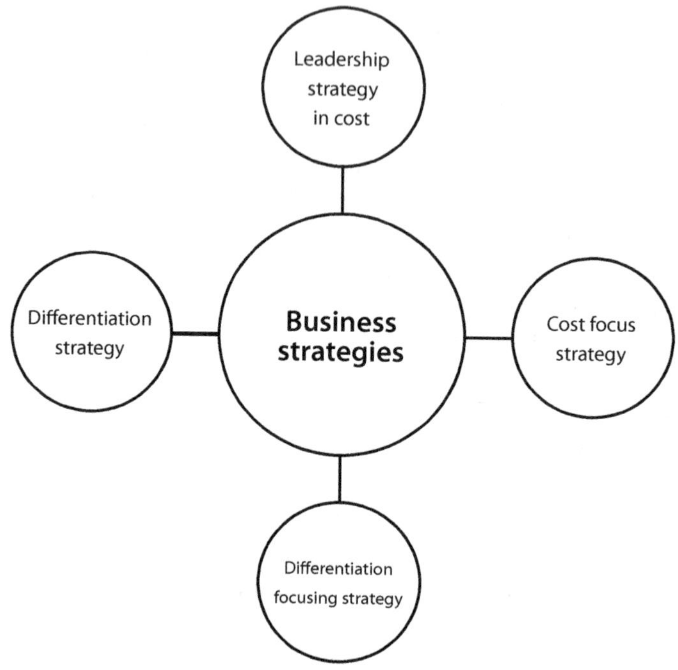

Fig. 7.2. Competitive strategies of M. Porter

in a narrow market segment. Thus, in order to achieve competitive advantages with a high-tech organization (for example, a helicopter manufacturing company), whose products usually compete in a narrow segment of the market, the most relevant strategies are focusing on costs and differentiation. Let us consider them in more detail.

The idea of the strategy of focusing on costs, otherwise called cost leadership, is to reduce the cost of the product by:

- high level of performance;
- high capacity utilization;
- using low-cost resources;
- lean production measures;
- effective use of modern technologies in the production process;
- access to the most effective distribution channels.

Applying one or more of these measures will allow the organization to follow a strategy of focusing on costs and thus create a competitive advantage, which in turn leads to a competitive advantage over other high-tech organizations.

The strategy of focusing on differentiation is to differentiate within one or a small number of target market segments, which is consistent with the principles of a modern high-tech organization. This classic marketing strategy allows an organization to establish in a niche market segment due to special knowledge and other ways to increase product value for customers.

As a rule, the products of high-tech organizations meet the special needs of customers, which may vary. Due to differentiation, organizations are able to adapt a particular product to individual customer needs and thus achieve priorities in development.

The global competitive advantage of an organization means its ability to produce and market products on international markets that exceed not only products of domestic firms, but also those of foreign ones. Therefore, the system of managing the process of achieving competitive advantage by an organization implies making effective management decisions, since at present they are the basis for creating promising products (products of advanced development), as well as forming its competitive advantages.

Effective management of an organization, its projects and programs is of fundamental importance when making decisions related to its successful operation and development. The concept of sustainability of such management is closely related to issues of effective management. By sustainable we mean management that ensures the achievement of goals and the fulfillment of all of an organization's tasks in the face of the manifestation of random factors affecting the course of their implementation.

Random factors (or risk factors) affecting the financial and economic activities of an organization can be both internal and external in nature. Sustainability of an organization's management implies the independence of the results of projects and programs from the impact of primarily internal factors. Such risk factors may include aging of personnel, low qualification of personnel, and depreciation of fixed assets. External risk factors are difficult to predict, and an organization's management has to plan its activities based on official forecast values of exchange rates, inflation, and other macroeconomic indicators that are relevant at the time of planning, as well as take into account existing interstate trade relations at the time of planning. These factors are of considerable importance in the current geopolitical environment.

The basis of any organization's activity is based on business processes that ensure the implementation of all types of activities related to the production of goods or services. For this reason, production process management mechanisms are being developed that are integrated into an organization's business process management system, which considers production processes as special resources of the organization that are continuously adapted to constant changes.

One of the most important principles of business process management is the ability to dynamically rebuild business process models by participants using modern software. Today on the Russian market you can find a large number of software products that help simplify the process description of an organization and conduct further business modeling.

Choosing business modeling tools is an important task. There are many criteria and attributes that can be used to evaluate business modeling tools. When choosing a software product for describing business processes, it is necessary to compare different products.

Let us describe the choice of business modeling tools for an organization that meets the ISO 9001:2000 standard, an international standard that regulates the requirements for a quality management system. For example, let us consider the activities of an organization in the field of purchasing materials and purchased components. The effectiveness of this procurement process is an important tool for managing the company's resources and costs. The ISO 9001: 2000 standard regulates procurement activities in clause 7.4. According to the requirements of ISO 9001: 2000 (clause 7.4. Purchasing) an organization must perform a number of actions when selecting a product and supplier (table. 7.1).

Table 7.1. Initial data for selecting a business modeling information system in the procurement sector

ISO standard point	Standard text	Meaning
7.4.1. Procurement process	«The organization must ensure that the purchased products meet the established procurement requirements. The type and degree of management applied to the supplier and the purchased product should depend on its impact on the subsequent stages of the product life cycle or the finished product»	Requirements for purchased products must be formed and the importance of each criterion must be evaluated
7.4.1. Procurement process	«The organization must evaluate and select suppliers based on their ability to deliver products in accordance with the organization's requirements. Criteria for selection, evaluation, and re-evaluation should be developed. Records of the evaluation results and any necessary actions resulting from the evaluation must be maintained (4.2.4)»	Criteria must be set not only for the product, but also for the product supplier itself. The supplier selection process must be documented. The supplier must be evaluated periodically
7.4.2. Information on procurement	«Purchasing information should describe the products ordered, including where necessary: a) requirements for approval of products, procedures, processes and equipment; b) requirements for qualification of personnel; C) requirements for the quality management system»	If necessary, the customer can set the developer criteria for: a) product development and testing process; b) qualification requirements for personnel who participate in product development; c) requirements for the developer's organization management system
7.4.2. Information on procurement	«The organization must ensure that the established procurement requirements are adequate before they are communicated to the supplier»	Requirements must be formed, should not contradict each other, and only then be presented to the supplier

ISO standard point	Standard text	Meaning
7.4.3. Verification of purchased products	«The organization must develop and implement controls or other activities necessary to ensure that the purchased product meets the established procurement requirements»	Purchased products must be checked for compliance with the supplier's requirements
7.4.3. Verification of purchased products	«If the organization or its consumer intend to perform verification at the supplier's enterprise, the organization must set in the procurement information the intended verification measures and the procedure for the production of products from the supplier»	Verification of purchased products and processes that create them can be performed at the supplier. To do this, you must create requirements for the supplier and check criteria for it

The business modeling system is designed to automate the management of internal regulatory documentation of an organization in terms of describing the processes and procedures of its main activities, and automatically generating reports (regulations). These regulations generated by the system must fully comply with the documents required by the international standard ISO 9001: 2000.:

- Quality Policy and Goals, Quality Guide;
- process regulations, procedure regulations;
- job descriptions, regulations on units;
- quality records;
- audit results, corrective, preventive actions.

To describe the Quality Policy and Goals, a special section "Management" is provided in the business modeling system, which is designed to store all information about the organization's management activities. As part of developing a quality management system, you can store quality Policies and Goals in this section. The creation of the Quality Guide is the final stage of the documentation creation project, it contains all the information about the system. When implementing the principle of continuous improvement, the quality guide may change many times. To significantly reduce the complexity of changes to this and related documents, a special static report is generated – the "quality Manual", which collects all the necessary information for such a document. The quality manual can be created in the form of a complex, composite report on the quality management system (QMS) containing both a "stationary" description of the system and attached, dynamically displayed reports, for example: quality policy, goals (report) in the field of quality, year, organizational structure, process landscape and other types of information that may change over time and their content is directly related to the management model.

The requirement of the ISO 9001:2000 standard for implementing a process approach in an organization is implemented in the business modeling system as part of creating a process model. The process model for different categories of users can be presented in the most convenient form. Executors get rules of procedures that they participate in; process owners get rules of processes that allow them to see and manage the entire process without going into unnecessary details. For all employees, a process landscape is formed, which allows you to see the organization's activities even more widely, and to understand the relationships between processes.

The business modeling system should have a tool for creating job descriptions, since the employee's job responsibilities are determined by his role and scope of activity. The position of the division depends on its participation in the processes and procedures of the enterprise. The possibility of simple dynamic updating allows you to re-issue outdated documents without much cost, and in the case of using electronic storage of information, the speed of document changes can coincide with the speed of changes in activity.

Department regulations and job descriptions based on actual activities allow managers to clearly see and evaluate the functional workload of employees. Based on the system of criteria, you can make a comprehensive assessment of software products for modeling business processes. For example, the software package of business modeling in the field of procurement should have the following features (table. 7.2).

Table 7.2. Features of the software package of business modeling
in the field of procurement

ISO Standard Point	Acceptable implementation in a business modeling system
7.4.1. Procurement process	The product evaluation criteria are proposed in the comparison table template. A demo version is provided in which all parameters of the full-featured version can be evaluated /verified
7.4.1. Procurement process	The evaluation criteria for the product in the comparative table template are proposed
7.4.2. Information on procurement	Creating requirements in the form of a comparative table template
7.4.3. Verification of purchased products	The mechanism for permanent monitoring

The description of the business process in the business modeling system is based on the following scheme:

1. business process goals that change over time. Goals are determined based on the requirements for the results of the process;
2. process result: objects (material, information) that are the result of the business process, consumed by other business processes or external customers in relation to the organization;
3. process owner: an official who has at his disposal personnel, infrastructure, etc., and also manages the business process and is responsible for its results and effectiveness;
4. executors of the process: officials who directly carry out the operations that make up the business process;
5. process documentation that regulates the execution of the process;
6. process diagram: a graphical diagram drawn up in a specific simulation notation;

7. interaction with other processes and the external environment: communication of inputs and outputs of the described process with inputs and outputs of other processes;
8. process execution organization: algorithms and activities that make up the business process;
9. process indicators: quantitative and / or qualitative parameters that characterize the business process and its result. There are 3 groups of indicators: process performance indicators, process product indicators, and process customer satisfaction indicators.

A business modeling system should describe a set of business processes of an organization in the form of some model, on the basis of which it is possible to consider various options to manage it. In a market environment, the content of an organization's activities includes independent provision of resources and sales of products and services. However, today this representation of the subject of an organization's activities is incomplete, since it does not reflect the special component that is the definition of the goals of the activity and without which it is impossible to implement all other processes. Therefore, the current model of an organization's activities in a changing market environment should be presented as follows (Fig. 7.3).

Fig. 7.3. An organization's model an in a changing market environment

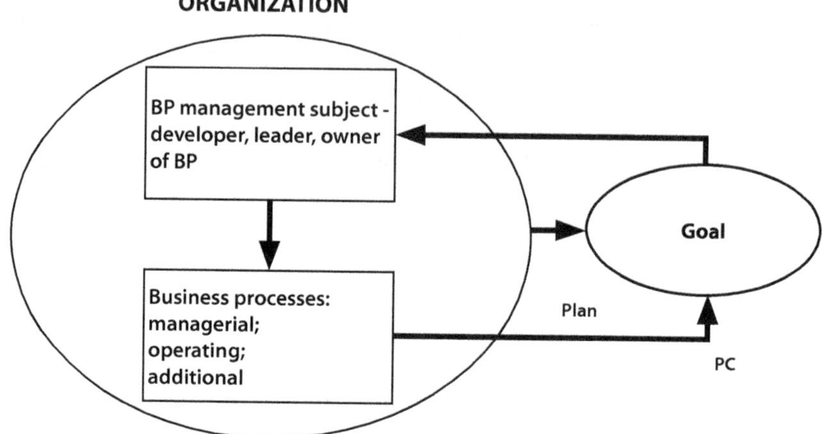

Fig. 7.4. Cyber-economic model of business process management
in an organization

In addition, under the new conditions, activity as an object of an organization expands in accordance with the cybernetic model (Fig. 7.3), and the range of organization tasks in its most general form includes:

1. organization of activities to develop the company's goals in a changing environment;
2. providing resources, converting them into performance results and implementing them;
3. organization of analysis of achievement of activity goals as a basis for a new cycle of setting goals.

If you imagine the activities of an enterprise on the base of the business processes, then their management based on the business modeling system models is possible due to the principles of cybereconomics. Let us imagine schematically a generalized cyber-economic model for managing any business process (Fig. 7.4) and describe its elements.

To initiate the management and improvement (reengineering) of business processes, a center for analyzing the quality of existing business processes and substantiating the processes that will become candidates for reengineering should function as part of the organization's management entity. Assessing the quality of business processes requires appropriate criteria. A general quality criterion can be any indicator that expresses the conformity of the results of a business process with its goals or functions. However, business process reengineering can be aimed at improving a separate partial indicator of its quality, if this does not contradict the general criterion. Systematic inconsistency of business process results with the set goals may indicate the need to reengineer the goal-setting process itself, that is, the process of developing goals. The goals of an organization are implemented by performing certain actions or functions. Management functions are areas or types of management activities based on division and cooperation in management, characterized by separate sets of tasks and performed by special techniques and methods. Any management function includes some information, its transformation, decision-making, giving it a certain form and bringing it to the performers.

Depending on the characteristics of the managed object, the General management functions themselves can be presented as complex, including various sub-functions.

For example, the planning function may consist of a sub-function of strategic or current planning, business planning, and sometimes forecasting.

An organization's function can be implemented in the form of:

- organizational design, i.e. creating projects or documented organizational solutions;
- organizational management as a form of direct development and transfer of organizational decisions without organizational design.

The control function is carried out through the implementation of the accounting sub-function. It consists in recording facts, performing tasks and sub-functions of control, as well as comparing actual data and indicators with the planned ones. The accounting sub-function itself is implemented in at least two forms: accounting, which represents a special professional activity as part of management, and management accounting, carried out by all departments of organizations based on their own tasks.

The structure of the regulatory function includes a sub-function of analysis, which consists in assessing the discrepancy between the planned and actual data that appear in the control process, and making a managerial decision if it is necessary to eliminate the inconsistency between the tasks and the results of their implementation.

Of course, if it is possible to make management decisions in advance, the management entity uses all the necessary functions, in emergency conditions-the composition of management functions is reduced. It should be noted that the function of an organization to some extent is carried out with all management options, since the order of interaction of the elements of the executive system can be determined only within an organization. In the absence of organizational support for the management decision, the organization's function is manifested at least in the form of self-organization.

CONCLUSION

As a result of the conducted research, it is established that in the conditions of digital transformation, dynamic development of science and technology, organizations seeking to ensure sustainable economic development should focus on creating products and services with unique consumer properties that could lead them to the path of advanced development. In this regard, the achievement of advanced development is impossible without the use of modern technologies of Industry 4.0, which make it possible to find the most effective design solutions using the intelligent analysis of big data both in shaping the techno-economic image of promising products, and in determining the ways to achieve their price and non-price competitive advantages.

The monograph develops economic tools aimed at stimulating the achievement of high competitive advantages of the organization on the basis of personification and customization of products, the manufacture of products with techno-economic characteristics that can meet the needs at a high level, the use of flexible production systems and the implementation of digital transformation of the main technological processes.

Methods of digital design and modeling, design for a given cost and competitiveness with the use of digital twins of products, organization, competitive market environment are proposed. The proposed methods for achieving high product competitiveness consider the organizational and economic aspects of their creation and will allow to manage all the processes from determining the image of a product to the moment of its replacement with a product with a higher consumer value with more advanced consumer properties, which will eventually allow to maintain the competitiveness of products at a high level.

Given the fact that the implementation of innovative projects is related to exploratory scientific research and the solution of new problems, there is an urgent task of developing a methodological apparatus for optimizing costs at various stages of the life cycle of innovative projects to create high-tech products, which will significantly reduce the cost of production and increase the competitiveness of products. In this regard, a relevant methodology has been proposed that allows to determine the optimal level of funding for an innovative project to create promising products, providing a given level of project funding efficiency. In addition, the algorithms of the methodology enable finding the optimal distribution of funding for various stages of the life cycle of an innovative project.

Since the sustainable economic development of an organization directly depends on the effective use of innovative technologies for the manufacture of existing and new products, as well as on the ways to form competitive advantages of products and services, it is necessary to determine the image of promising products based on market analysis and consumer expectations.

In this regard, a methodology has been developed for quantitatively assessing the impact of digital technologies on the product value chain, which makes it possible to absolutely and relatively assess the impact of digital technologies on the value chain of promising products. It is shown that this mathematical model can be used as a basis for the methodological apparatus for assessing the impact of digital technologies.

Digital models of products and production and technological processes are proposed, focused on the optimal management of the creation and manufacture of a product using modern intelligent cyber-physical and cyber-economic systems that create an integrated digital platform for supporting effective solutions in managing the life cycle of a product with appropriate information support.

Based on the study of a number of factors necessary for making an effective managerial decision on the need to make certain changes in the techno-economic characteristics of a high-tech product, a method is proposed for a comprehensive assessment of the effectiveness of product renewal when it is created for given techno-economic parameters, which will increase the degree of mutual understanding. between divisions of the organization, and will also help to form an approach among managers that meets the needs in the development of products of the future, which in turn will allow the organization to achieve advanced development, in which it can become the most competitive in the existing market or create a new market.

BIBLIOGRAPHY

1. *Angang, Hu.* Embracing China's «New Normal» / Hu Angang // Foreign Affairs. – 2015. – № 3. – P. 5–11.
2. *Baden-Fuller, C.* Business models as models / C. Baden-Fuller, M.S. Morgan // Long Range Planning. – 2010. – № 43. – P. 154–156.
3. *Boisoit, M.* Knowledge Assets. Securing Competitive Advantage in the Information Economy / M. Boisoit. – Oxford: Oxford University Press, 1998.
4. *Boginsky, A.I., Chursin, A.A.* Optimization of Production Processes by Means of Digital Models // Russian Engineering Research, (2020), 40 (5), pp. 396-399.
5. *Boginsky, A.I., Zelentsova, L.S., Tikhonov, A.I.* Intelligent Monitoring as the Basis for Improving the Competitiveness of High-Tech Production // Lecture Notes in Networks and Systems, (2020) 115, pp. 436-443.
6. *Boginsky, A.I., Chursin, A.A.* Optimizing Product Cost // Russian Engineering Research, (2019) 39 (11), pp. 940-943.
7. *Boginskiy, A.I., Chursin, A.A., Nesterov, E.A., Tyulin, A.E.* – Assessing the competitiveness of production of a high-tech corporation to ensure its advanced development // Journal of Advanced Research in Dynamical and Control Systems. 2019. – T. 11. – № 11 Special Issue. – P 73-81.
8. *Chen, H.C.* Knowledge and Development: a Cross-Section Approach / H.C. Chen, C.J. Dahlman // World Bank Policy: Research Working Paper. – August 2004. – № 3366.
9. *Collis, D.* Competing on Resources: Strategy in the 1990s / D. Collis, C.A. Montgomery // Harvard Business Review. – 1995. – № 73. – P. 118–128.
10. *Conte, T.A.* Framework for Business Models in Business Value Networks / T. Conte// White Paper. – London: Institute of Information Systems and Management (IISM), September 8, 2008. – № 001-08. – P. 5-7.

11. *Crosby, P.* Quality Is Free: The Art of Making Quality Certain: How to Manage Quality – So That It Becomes A Source of Profit for Your Business / P. Crosby. – N. Y.: McGraw-Hill Companies, 1979. – 309 p.

12. *Cukier, K.* Survey: Patents and Technology / K. Cukier // The Economist. – 2005, October 22. – P. 3–20.

13. *Davenport, T.* et al. Successful Knowledge Management Projects / T. Davenport et al. // Sloan Management Review. – February 1998. – № 17 (8). – P. 43–57.

14. *Davenport, T.* Working Knowledge / T.Davenport, L. Prusak. – Boston: Harvard Business School Press, 2000.

15. *Davenport, T.H.* Improving knowledge work processes / T.H. Davenport, S.L. Jarvenpaa, M.S. Beers // MIT Sloan Management Review. – 1996. – № 8. – Vol. 1. – P. 53–65.

16. *Deming, W.E.* Quality Productivity and Competitive Position / W.E. Deming. – Massachusetts: Massachusetts Inst Technology, 1982. – 373 p.

17. *Department* of Defense Fiscal Year 2016. President's Budget Submission, 2015.

18. *Dörnbusch, R. Macroeconomics* / R.Dornbusch, S. Fischer, R. Startz. – N. Y.: McGraw-Hill Education, 2013. – 672 p.

19. *Drucker, P.* Post Capitalist Society / P. Drucker. – Oxford: Butterworth & Heinemann, 1993. – 232 p.

20. *Drucker, P.* The Age of Discontinuity/ P.Drucker. – Piscataway, New Jersey: Transaction Publishers, 1969. – 436 p.

21. *Drucker, P.* The Next Information Revolution / P.Drucker// Forbes ASAP. – 1998. – № 24. – P. 57–69.

22. *Edvinsson, L.* Intellectual Capital: Realizing Your Company's True Value by Finding Its Hidden Roots / L.Edvinsson, M. Malone. – N. Y.: HarperCollins Publishers, 1997.

23. *Feigenbaum, A.* Quality control: principles, practice, and administration / A. Feigenbaum. – N.Y.: McGraw-Hill, 1951. – 443 p.

24. *Firer, S.* Intellectual Capital and Traditional Measures of Corporate Performance / S.Firer, M. Williams. – N.Y., 2001.

25. *Flanagan, R.J.* Globalization and Labor Conditions: Working Conditions and Worker Rights in a Global Economy / R.J.Flanagan. – Oxford: Oxford University Press, 2006. – 272 p.

26. *Frascati Manual 2002:* Proposed Standard Practice for Surveys on Research and Experimental Development // OECD, 2002.

27. *FUTRON'S 2014 Space Competitiveness Index.* – Washington: Futron Corporation, 2014. – 180 p.

28. *Garicano, L.* Organization and Inequality in a Knowledge Economy / L. Garicano, E. Rossi-Hansberg // NBER Working Paper. – June 2005. – № 11458. – P. 16–25.

29. *Garvin, D.A.* Building a Learning Organization / D.A. Garvin // Harvard Business Review. – July–August 1993. – № 18 (7). – P. 78–91.

30. *Gibbons, M.* The New Production of Knowledge: The Dynamics of Science and Research in Contemporary Societies / M. Gibbons, C. Limoges, H. Nowotny, S. Schwartzman, P. Scott, M. Trow. – L.: Sage Publications, 1994.

31. *Granovetter, M.* Social Structure and Network Analysis / M. Granovetter. – Thousand Oaks: SAGE Publications Ltd, 1982.

32. *Grant, R.* Prospering in Dynamically-competitive Environments: Organizational Capability as Knowledge Integration / R. Grant // Organization Science. – July 1996. – № 4 (7). – P. 375–387.

33. *Hamel, G.* Competing for the Future / G. Hamel, C.K. Prahalad. – Boston: Harvard Business School Press, 1994.

34. *Hammer, M.* Reengineering the Corporation: A Manifesto for Business Revolution / M.Hammer, J. Champy. – N. Y.: Harper Business, 1993. – 272 p.

35. *Ishikawa, K.* What is Total Quality Control? The Japanese Way Hardcover / K. Ishikawa; translated by J.Lu. David. – Prentice Hall, 1988. – 240 p.

36. *Kay, J.* Strategy and the illusions of grand designs, Mastering Strategy / J. Kay // Financial Times. – 1999. – № 15. – P. 2–4.

37. *Kim and Mauborgne.* Blue Ocean Strategy. – Harvard: Harvard Business School Press, 2005.

38. *Luthy, D.* Intellectual Capital and its Measurement / D.Luthy // Asian Pacific Interdisciplinary Research in Accounting Conference (APIRA). – Osaka (Japan), 1998. – P. 115–119.

39. *Machlup, F.* The Economics of Information and Human Capital / F. Machlup // Knowledge, its Creation, Distribution, and

Economic Significance. – N.J.; Princeton: Princeton University Press, 1984. – № 3. – P. 67–78.

40. *Nesterov E.A., Tuylin A.E., Boginskiy A.I., Chursin A.A.* Method for assessing the economic sustainability of a high-tech corporation // IOP Conference Series: Materials Science and Engineering. Krasnoyarsk Science and Technology City Hall of the Russian Union of Scientific and Engineering Associations. Krasnoyarsk, Russia, 2020. P. 42016.

41. *Osel, Roger R. and Wright Robert V.L.* Allocating resources: How to Do It in Multi-Industry Corporations. Handbook of Business Problem Solving. New York: McGrow-Hill. 1980.

42. *Osterwalder, A.* Business Model Generation: A Handbook for Visionaries, Game Changers and Challengers / A. Osterwalder, Y. Pigneur. – Hoboken, New Jersey: John Wiley & Sons, Inc., 2010.

43. *Ouchi, W.G.* Theory Z: How American business can meet the Japanese challenge / W.G. Ouchi. – Boston: Addison-Wesley Pub, 1981. – 192 p.

44. *Penrose E.* The theory of growth of the firm. New York, M.E. Sharpe, 1980. – 265 p. 21.

45. *Penrose, E.* The theory of growth of the firm / E.Penrose. – N.Y., 1959.

46. *Porter, M.E.* Competitive Strategy: Techniques for Analyzing Industries and Competitors / M.E. Porter. – N.Y.: The Free Press, 1980. – 397 p.

47. *Porter, M.E.* Measuring the Ideas Production Function: Evidence from International Patent Output / M.E. Porter, S. Stern // NBER Working Paper. – Camb., MA: National Bureau of Economic Research, 2000. – № 7891. – P. 46-53.

48. *Schein, E.H.* Culture: the missing concept in organizational studies / E.H. Schein // Administrative Science Quarterly. – 1996. – Vol. 41. – № 2. – P. 229–240.

49. *Selznick, P.* Leadership in Administration / P. Selznick. – N.Y.: Harper, 1957.

50. *Senge, P.M.* The Fifth Discipline Fieldbook: Strategies and Tools for Building a Learning Organization Paperback / P.M. Senge. – N.Y.: Doubleday, 1994. – 593 p.

51. *Shewhart, W.A.* Economic control of quality of manufactured product / W.A. Shewhart. – N. Y.: D. Van Nostrand Company, 1931.

52. *Sloan, A.* My Years with General Motors / A. Sloan. – N.Y.: Doubleday, 1964. – 472 p.

53. *Smith, A.* The Wealth of Nations / A.Smith. – North Charleston: Create Space Independent Publishing Platform, 2014. – 524 p.

54. *Solow, R.* Technical change and the aggregate production function / R. Solow // The Review of Economics and Statistics. – 1957. – Vol. 39. – № 3. – P. 312–320.

55. *Tallman, S.* Business Models and Multinational Firm / S. Tallman// Multidisciplinary Insights from New AIB Fellows Research in Global Strategic Management. – 2014. – Vol. 16. – P. 115–138.

56. *Teece, D.J.* Business Models, Business Strategy and Innovation / D.J. Teece // Long Range Planning. – 2010. – № 43. – P. 172–194.

57. *The Seventh Framework Programme (FP7).* Brussel. 2007.

58. *The Sixth Framework Programme in brief.* Brussel. December 2002.

59. *The State of Human Capital 2012:* False Summit. Why the Human Capital Function Still Has Far to Go: A Report by McKinsey & Company and The Conference Board. – N.Y., 2012.

60. *Tikhonov, A.I., Sazonov, A.A., Boginsky, A.I.* Planning, Development, and Quality Systems of Helicopters Production in Russia // Lecture Notes in Networks and Systems, (2020)/ –115, pp. 650-662.

61. *Tyulin, A.E., Boginsky, A.I., Chursin, A.A.* Key Technological Competencies and the Economic Efficiency of Space Technology // Russian Engineering Research, (2020), 40 (1), pp. 6-10.

62. *Unleashing America's Research & Innovation Enterprise.* – Cambridge, Massachusetts: American Academy of Arts and Sciences, 2013.

63. *Wernerfelt, B.* A Resource-based view of the firm / B. Wernerfelt // Strategic Management Journal. – 1984. – Vol. 5. – № 2. – P. 171-180.

64. *Akerman, Ye.N.* Transformation of models of innovative development on the way to openness of innovative systems /

Akerman, Ye.N., Yu.S. Burets // Tomsk state university journal – 2014. – № 378. – P. 178-183.

65. *Asaul, A.N.* The modernization of the economy based on technological innovation / A. Asaul, Karpov B.M., Perevyazkin V.B., Starovoytov M.K. – SPb: ANO IPEV, 2008. – 606 p.

66. *Bely, E.* Integrated structures in the modern economy: the essence and trends of development / E.M. Bely, Rozhkova E.V, Tyulin A.E. // Fundamental research. – № 6. – 2013. – P. 1482-1484.

67. *Bely, E.* Statistical tools to solve the problems of integration of industrial enterprises / E.M. Bely, A.E. Tyulin E.M. Bely, A.E. Tyulin // materials of the 1st International Conference on the formation of the main directions of development of modern statistics and econometrics. Orenburg, September 26-28, 2013 – vol. 1. – Orenburg, 2013. – P. 204-211.

68. *The Open Future* // Thomson Reuters. – 2015. – P. 13–35.

69. *Vanyurikhin, G.* The economy of space activities: Monograph / Vanyurikhin, G.I., Paison, D.B., Makarov, Yu.N. and others; under the scientific. ed. of the Doctor of technical sciences, prof. G.G. Raikunova. – Moscow: FIZMATLIT, 2013. – 600 p.

70. *Vikhansky, O.* Management: The Textbook / Vikhansky, O.S., Naumov, A I., – 4 ed., revised. and extras. -M.: Economist, 2006. -670 P.,

71. *Womack, J.* Lean Production / Womack, J., Jones, D. – M.: Alpina-Publisher, 2004.

72. *Gates, B.* Business @ the speed of thought: English translation /B. Gates. -M.: Eksmo-press, 2001. – 480 p.

73. *Gitelman, L.D.* Centers of competencies – the progressive form of organization of innovative activity / Gitelman, L.D., Kozhevnikov, M.V. // Innovation. – 2013. – № 10 (180). – P. 92–98.

74. *Ivanter, V.* The innovative and technological development of economy of Russia / Ivanter, V. V., ed. Ivanter, V. V. – Max Press, 2005. – 592 p.

75. *Ilyin, V.* Intellectual resources as an innovative development factor / Ilyin, V.A., Gulin, K.A., Uskova T.V. // facts, trends, forecast. – 2010. – №. 3. – T. 11. – P. 14-25.

76. *Castells, M.* Information Age: economy, society and culture: English translation / Castells, M.– M.: HSE, 2000. – 608 p.

77. *Kleiner, G.B.* The resource theory of the system organization of Economics/ Kleiner, G.B./Russian Management Journal. 2011. –№. 3. – P. 3-28.

78. *Maurik, J.* Effective strategist: English translation / John van Maurik. – M.: INFRA-M, 2002. – 208 p.

79. *Mintzberg, H., Ahlstrand, B., Lampel, J.* Schools of strategies / trans. from English. under the editorship of Yu. Kapturevsky. SPb.: Peter, 2001.

80. *Prahalad, C.K.* Core competence of the Corporation / C.K.Prahalad, G. Hamel // H.Mintzberg, J.Quinn, S.Ghoshal, Strategic process: trans. from English. under the editorship of Yu. Kapturevsky. SPb.: Peter, 2001.

81. *Seven notes of management.* Manager's Handbook / ed. V. Kondratiev – 7 Ed., revised. and extras. – M.: Eksmo, 2008. – 976 p.

82. *Soifer, V.* The development of a competence centre in the field of aerospace and geo-information technologies / V. Soifer, E.Shakhmatov // Herald of the Nizhny Novgorod University after N.I.Lobachevsky. – 2007. – №. 2. – p. 41-48.

83. *Stewart, T.A.* Intellectual capital. A new source of wealth for the organization: trans. from English by V. Nozdrina / T.A. Stewart. – M.: Pokoleniye, 2007. – 368 p.

84. *Suprun, V.* Intellectual capital: The main factor of competitiveness of economy in the XXI century / V.Suprun. – M.: Komkniga, 2006. – 192 p.

85. *Taylor, F.W.* The principles of scientific management / F.W.Taylor. – M.: Dashkov and co., 2008.

86. *Teece, D.J.* Dynamic capabilities and Strategic Management / D.J. Teece, G. Pisano, A. Shuen // Herald of St. Petersburg State University. "Management". – 2003. – № 4. – P. 133-171.

87. *Thompson, A.A.* Strategic management: Concepts and situations for analysis / trans. from English / A.A. Thompson, A.Strickland. – 12 ed. – M.: Williams Ed. House, 2006. – 928 p.

88. *Harrington, J.* Knowledge management excellence. / J. Harrington, F. Vole. – M.: Standards and quality, 2008.

89. *Hodgson, G.* Socio-economic Consequences of the Advance of Complexity and Knowledge / G.Hodgson // Questions of economy. – 2001. – №. 8. – P. 32-46.

90. *Hodgson, G.* Economic theory and institutions: The Manifesto of modern institutional Economics / G.Hodgson.. – M.: Delo, 2003. – 464 p.

91. *Tsvetkov, V.* Competencies and competitiveness of staff / V.Tsvetkov, K.Pushkareva // International journal of applied and fundamental research. – 2010. – №. 1. – P. 85-86.

92. *Chursin, A., Tyulin, A.* Competence management and competitive product development: Concept and implications for practice. Springer International Publishing. – 2018. – P. 1-241. DOI: 10.1007/978-3-319-75085-9

93. *Shinkevich, A.* Managerial innovations as a factor of productivity growth / A. Shinkevich, D. Sultanova, R.Burganov // Bulletin of the Kazan University. -2013. -№. 24. – Vol. 16. – P. 217-220.

94. *Shiryaeva, K.* Current state of space-rocket industry of Russia / K. Shiryaeva, Yu.Anishchenko // Actual problems of aviation and cosmonautics. – 2014. – №. 10. – T. 2. – P. 48-49.

95. *Shifrin, M.* Strategic management / M.Shifrin. – Spb: Peter, 2006.

96. *Schoenberg, R.* Japanese methods of business management / R. Schoenberg. – M.: Economica, 1988. – 251 p.

97. *Schumpeter, J.A.* Capitalism, Socialism and Democracy: trans. from English / J.A. 149. Schumpeter; foreword and the general editorship of V. Avtonomov. – M.: Economica, 1988. – 540 p.

TABLE OF CONTENTS